STRUCTURAL OPTIMIZATION

RECENT DEVELOPMENTS AND APPLICATIONS

Sponsored by
The Structural Division
of the American Society of Civil Engineers

Prepared by
The Committee on Optimization
of the Committee on Electronic Computation

Edited by Ovadia E. Lev

Published by the
American Society of Civil Engineers
345 East 47th Street
New York, New York 10017

TA
658.2
.S758

Copyright © 1981 by the American Society of Civil Engineers,
All Rights Reserved.
Library of Congress Catalog Card No. 81-69232
ISBN 0-87262-281-9
Manufactured in the United States of America.

FOREWORD

This state-of-the-art report is the result of a joint effort, over a period of approximately 18 months, undertaken by an ad-hoc committee which was formed for this purpose by the Committee on Optimization of the Committee on Electronic Computation of the Structural Division of ASCE. The first draft of the report was completed in June of 1980.

The following is a list of committee member who directly and actively authored the various chapters of the report.

    Raphael T. Haftka, Illinois Institute of Technology
    Edward J. Haug, The University of Iowa, Iowa City
    Harry L. Jones, Texas A & M University
    Narbey Khachaturian, University of Illinois, Urbana
    Narendra S. Khot, Air Force Flight Dynamics Laboratory
    Ovadia E. Lev, Merritt CASES, Inc.
    Fred Moses, Case Western Reserve University
    William R. Spillers, Rensselaer Polytechnic Institute
    H. Randolph Thomas, Jr., Pennsylvania State University
    William A. Thornton, Cives Steel Inc.
    Garret N. Vanderplaats, Naval Post Graduate School

Other members of the committee who interacted and contributed material to the report were:

    Jasbir S. Arora, University of Iowa, Iowa City
    Lewis P. Felton, University of California, Los Angeles
    V. Kalyanaraman, University of Kentucky, Lexington
    Uri Kirsch, Technion, Israel
    C. Wayne Martin, The University of Nebraska, Lincoln
    S. Ramamurthy, CONRAIL, Pennsylvania
    George I. N. Rozvany, Monash University, Australia
    Vipperla B. Venkayya, Air Force Flight Dynamics Laboratory

Many researchers and engineers contributed to this report by responding to calls for information and answering the survey questionnaire. Their contribution is gratefully acknowledged. Special thanks are due to Professor Judith S. Liebman for her contribution to the list of books on optimization.

The American Society of Civil Engineers would like especially to thank Merritt CASES, Inc. for contributing the time, clerical, and administrative work which made this report possible. The support and encouragement of the editor by Dr. J. L. Merritt is particularly appreciated. The ASCE would also like to thank Pat Wheatley of Boeing Computer Services Company for the complete final typing of this report.

                                                           Ovadia E. Lev, Editor

TABLE OF CONTENTS

PART I: INTRODUCTION

                                                                          Page

1. INTRODUCTION 7

    1.1 OBJECTIVES 7
    1.2 OVERVIEW 7
    1.3 SCOPE AND ORGANIZATION 9

2. STRUCTURAL OPTIMIZATION 12

    2.1 TERMINOLOGY 12
    2.2 THE MATHEMATICAL PROBLEM 12
    2.3 CLASSIFICATIONS 14
    2.4 METHODS OF OPTIMIZATION 14
    2.5 DEVELOPMENT AND LITERATURE 16
    2.6 REFERENCES 17

PART II: NUMERICAL METHODS

3. INDUSTRIAL APPLICATIONS 21

    3.1 INTRODUCTION 21
    3.2 COMPONENT DESIGN 21
        3.2.1 Box Girder Design 22
        3.2.2 Welded Plate Girder 22
        3.2.3 Standardized Frames 23
        3.2.4 Reinforced Concrete Beams 23
        3.2.5 Shear Wall Design 24
        3.2.6 Prestressed Concrete Beams 24
        3.2.7 Connection Design 24
    3.3 MASS-PRODUCED STRUCTURAL SYSTEMS 24
        3.3.1 Grillages 25
        3.3.2 Transmission Towers 25
        3.3.3 Roof Trusses 25
        3.3.4 Prestressed Concrete Bridge Decks 26
    3.4 SPECIALIZED APPLICATIONS 26
        3.4.1 Design of Continuous Highway Bridge Girders 27
    3.5 SHAPE OPTIMIZATION 27
        3.5.1 Difficulties 28
        3.5.2 Topological Considerations 28
        3.5.3 Methods 28
        3.5.4 Applications 29
    3.6 REFERENCES 31

4. AEROSPACE APPLICATIONS 35

    4.1 HISTORICAL OVERVIEW 35

TABLE OF CONTENTS (continued)

Page

    4.2 APPLICATION LITERATURE REVIEW 36
        4.2.1 Stress and Displacement Constraints 36
        4.2.2 Buckling Constraints 42
        4.2.3 Natural Frequency Constraints 43
        4.2.4 Thermal Constraints 43
        4.2.5 Aeroelastic Constraints 43
        4.2.6 Reducing the Number of Analyses 44
    4.3 CONCLUSIONS 44
    4.4 REFERENCES 46

5. AN OUTLOOK FOR BASIC RESEARCH 57

    5.1 INTRODUCTION 57
    5.2 MATHEMATICAL PROGRAMMING ALGORITHMS 59
    5.3 STRUCTURAL OPTIMIZATION 59
    5.4 HARDWARE 61
    5.5 REFERENCES 63

6. CONCLUSIONS 64

    6.1 THE STATE-OF-THE-ART 64
        6.1.1 Algorithm Efficiency 65
        6.1.2 Fully Stressed Design (FSD) 66
    6.2 DESIGN APPLICATIONS 66
    6.3 PROJECTIONS AND FUTURE NEEDS 67
        6.3.1 Research 67
        6.3.2 Applications 68

7. FREQUENTLY REFERRED TO EXAMPLES 70

8. REFERENCES 93

    8.1 BOOKS ON STRUCTURAL OPTIMIZATION 93
    8.2 GENERAL BOOKS ON OPTIMIZATION 95
    8.3 PAPERS ON OPTIMIZATION................................. 102
    8.4 COMPUTER PROGRAMS ON STRUCTURAL OPTIMIZATION 145

PART III: ANALYTICAL METHODS

9. DISTRIBUTED PARAMETER STRUCTURES 148

    9.1 INTRODUCTION 148
        9.1.1 Literature Reviews 149
        9.1.2 Books 150
        9.1.3 Applicable Distributed Parameter Optimal 150
              Control Literature
    9.2 BUCKLING OF COLUMNS, PLATES AND SHELLS 152
        9.2.1 Column Buckling................................ 152
        9.2.2 Other Buckling Problems........................ 155

TABLE OF CONTENTS (continued)

	Page
9.3 VIBRATION OF BARS, BEAMS AND PLATES	156
9.3.1 Vibration of Bars	156
9.3.2 Vibration of Beams	157
9.3.3 Torsional Vibration of Shafts	158
9.3.4 Vibration of Plates	159
9.4 DEFLECTION AND COMPLIANCE OF BEAMS AND PLATES	160
9.4.1 Beams with Deflection Constraints	160
9.4.2 Plates with Deflection Constraints	162
9.4.3 Beams and Plates with Compliance Constraints	163
9.4.4 Miscellaneous Problems	164
9.5 DYNAMIC RESPONSE	164
9.5.1 Forced Steady State Oscillation	164
9.5.2 Transient Dynamic Response	165
9.5.3 Earthquake Structural Design	167
9.5.4 Miscellaneous Problems	168
9.6 SHAPE OPTIMAL DESIGN	168
9.6.1 Numerical Methods for Shape Optimization	168
9.6.2 Shape of Cross Section of Shafts in Torsion	169
9.6.3 Shapes of Holes in Planar Solids	170
9.6.4 Miscellaneous Shape Optimal Design Problems	170
9.6.5 Related Literature on Domain Optimization	171
9.7 MULTIPURPOSE STRUCTURES AND MISCELLANEOUS OPTIMAL DESIGN PROBLEMS	171
9.7.1 Multipurpose Structures	171
9.7.2 Special Problems	172
9.8 REFERENCES	175

APPENDICES

    A. ANALYTICAL METHODS IN STRUCTURAL OPTIMIZATION: AN UPDATE 197

    B. OPTIMALITY CRITERION APPROACH: RECENT ADVANCES 200

PART I
INTRODUCTION

1. INTRODUCTION

1.1 OBJECTIVES

The purpose of this report is to provide a current documentation of the state-of-the-art of structural optimization, emphasizing recent developments in the field (approximately between 1972 and 1980). Based on these developments, conclusions are drawn regarding present trends, future needs, and projections.

This report is intended to be used as reference material by engineers and researchers. In addition, it may be used as a guide for the students and scientists who are interested in but are not familiar with structural optimization. It is hoped that the examples of applications presented will encourage the use of optimization in the design process.

1.2 OVERVIEW

For over 20 years, researchers have been investigating applications of mathematical programming and computer-based search algorithms to the solution of optimum structural design problems. Structural optimization has been defined as designing and constructing a structure at the lowest cost, with the objective of fulfilling a well-defined purpose. Costs must also include safety, service life, maintenance, and future adaptability. Since all research and practice in structural engineering are presumably aimed towards such a goal, the activity known as structural optimization must be defined in a unique way. That is, the development and application of interactive or automated computer techniques for improving designs within well-defined costs and constraints.

In this definition of structural optimization, the computer becomes central as a tool for searching and sorting through the similar design concepts and proportioning the element details for the most economical design. In general, optimization arrives at a design which the engineer could equally as well have obtained if he were prepared to invest the time and money to directly search among the design alternatives. The principal advantage of the optimization methods should, therefore, be its saving in design time and cost. A further consideration is that an optimization design program can eliminate much of the input-output and the costly data handling effort while producing a proportioned structure which satisfies applicable codes. The review of vast amounts of structural analysis output can thus be avoided. In some cases, optimization procedures could be further programmed to produce drawings and material specifications.

In recent years, as computer hardware costs have decreased, the cost of software and programming have become a major factor in computer usage. Thus, one-of-a-kind structures, unless of major economic importance, are less likely to warrant specialized computer analysis and design. It is, therefore, not surprising that the major implementation of structural optimization work has been industrialized applications. That is, repetitive structures or elements of structures which are built in sufficient quantity that the cost of computer program development can be retrieved by savings in a large number of similar applications. These include such diverse examples as overhead cranes, standardized metal buildings, transmission towers, short and medium span highway bridges, and a variety of vehicles including aircraft, automobiles, and ships. These cases, which by their repetitive nature we may describe as industrialized structures, must have several additional features to warrant structural optimization. The structures must be sold in an economic climate which encourages competitive cost savings and further, structure cost must be a significant proportion of overall sale price which may include land, equipment, financing, and furnishings. In addition, the organization producing the design must have the capability of using computers and, if acting as a consultant, have the incentive to balance the increased computer costs with savings in the structure. All these conditions are ideally fulfilled by airplanes and automobiles. Indeed the aerospace industry has been the major investor in optimization and there are many indications that the auto industry is catching up.

While still regarded by many practicing engineers as primarily a research tool, the concept of structural optimization has been firmly established. The need to make optimization procedures an integral part of the design process is becoming clearer as more sophisticated and powerful computer analysis techniques continue to develop. Structural optimization seems to be progressing in two fronts. In the first, theoretical research is producing new algorithms and improved techniques of optimization. The sources of this progress continue to be analytical methods of structural optimization, general optimization theory, control theory and methods of operations research, as well as parallel developments in fields other than structural engineering. In the second front, simple optimization methods are finding increasing use within various phases of the design process. The steady decrease in the cost of computer usage, mainly due to recent hardware developments, is now making the use of blackbox optimization procedures quite attractive for the design office and the numerous analyses that are usually required are no longer found unacceptable. As opposed to earlier unrealistically optimistic projections that structural optimization might supplant the need for good designers, the current trend seems to indicate that experienced designers would benefit the most from this tool which, if used properly, would enable them to systematically examine and fine-tune tens or hundreds of design variants very efficiently. In this era of energy shortage and conservation needs, methods and applications of optimization will, no doubt, be developing as an indispensable module of computer-aided design.

To conclude this overview, the results of a recent survey conducted on a small but carefully selected sample of professionals, asking them for their opinions on structural optimization, are presented.

Of the 27 respondents, 11 were in engineering practice, 11 were engaged in research, and 5 indicated that they were both in practice and in research. The respondents in practice represented a wide range of activities. Their practice involved design of different types of structures including buildings, bridges, power plants, floating vessels, offshore structures, earth structures, and harbor facilities. The materials used were metals and structural concrete.

Most of those engaged in practice indicated that cost-saving was an important consideration in their design activities. There was no indication, however, that cost-saving was effected by the use of optimization techniques. Most did not have sufficient familiarity with these techniques to make use of them in practice. In general, practicing engineers make very little use of mathematical programming in design. One respondent indicated that he used the exhaustive search approach to determine the optimal values of the design variables.

The respondents in research were engaged in a wide spectrum of activities ranging from development of algorithms to optimization of specific structures. They agreed that optimization would provide an expeditious approach to design of structures, and that more work is required to obtain usable programs. One respondent indicated that a program of large capability was already being used in the design of cars.

The researchers generally thought that use of mathematical programming in design of structures is only limited to very specific situations, and is not widespread. They felt that in order to encourage use of optimization in design, it is necessary to develop better software, and educate the engineers in their use. Some thought that the present research is too theoretical.

The researchers also felt that we should demonstrate to the practicing engineers the usefulness of the methods of optimization, develop practical programs, and teach courses on optimization techniques.

The respondents in practice indicated that they are interested to know more about optimization if its usefulness can be demonstrated. Only one respondent indicated that he would not be interested.

While this sample is statistically small, it is believed to describe the present status of structural optimization as a design tool quite accurately.

1.3 SCOPE AND ORGANIZATION

Consistent with the objectives of this report, the review of recent literature is quite extensive. However, because optimization is pervasive and the number of recent works is so large, the coverage may not be exhaustive. Furthermore, in reporting the developments in the field, mention of some earlier works could not be avoided. The period covered by the report is approximately a decade (1970-1980), with emphasis on developments between 1972 and 1979.

No attempt was made to make this report a source of studying structural optimization. Instead, Chapter 2 is intended to guide the interested reader, who is not familiar with the subject, in tracing the original works, the theoretical bases, and the developments. Chapters 1 and 2 are general and constitute Part I of the report.

Research in structural optimization has and is being performed by two groups of scientists. One group emphasizes the analytical, conceptual aspect, the other emphasizing the algorithmical, numerical aspect. While both groups direct their efforts towards the same structural applications, a wide gap between them does exist. Clearly, the developments of numerical methods have been based on theoretical methods. However, because of the nature of engineering applications, at the practical design level, the progress in numerical methods has necessarily been slower. While this report concentrates on the numerical aspects and industrial applications of structural optimization (Part II), the analytical methods are also reviewed (Part III and Appendix A).

In Part II, Chapter 3 is dedicated to industrial applications. These are classified into component designs, mass-produced systems, and specialized applications. These applications focus primarily on the civil structural area, including buildings, bridges, and cranes. In order to present a representative sample of applications, requests for contributions were made through the ASCE Newsletter as well as through private correspondence. Of the many contributions which had been received, few were selected as representatives for applications and implementations. Section 3.5 of Chapter 3 reviews recent developments in shape optimization.

An extensive coverage is given to aerospace applications in Chapter 4. This is the area where structural optimization enjoyed the most support and development. After a review of early and recent literature, special problems with constraints on displacements, buckling, frequencies, etc., are discussed.

In the last decade, a significant progress has been made in the development and application of the optimality criterion methods. In recognition of this fact, Appendix B is dedicated exclusively to recent advances in these methods and their relations to other optimization techniques.

It should be remarked that major applications of structural optimization are currently in progress in the automobile industry. To the regret of the editor, attempts to obtain specific information on this subject have not been successful, possibly due to proprietorship.

Basic research and likely future trends in structural optimization are discussed in Chapter 5. Chapter 6 includes conclusions and recommendations.

A collection of frequently referred to examples is presented in Chapter 7 which also lists the references where each example is discussed. Chapter 8 includes three types of references: Books on Structural Optimization (8.1), Books on General Optimization Methods (8.2), and a large collection of papers (8.3). The references in Section 8.3 include some but not all of the references of Chapters 2, 3, 4, and 5. The cut-off year for the references in Section 8.3 is 1970. From these references a list of available computer programs was extracted and included in Section 8.4.

Part III on analytical methods includes an extensive review on optimization of distributed parameter structures. Other areas of analytical methods are included in a review and supplemented in Appendix A. These parts of the report will probably be only of general interest to practitioners who are interested in numerical applications.

2. STRUCTURAL OPTIMIZATION

This chapter is presented to introduce the terminology, problems, and historical developments of structural optimization to the interested reader who is not familiar with the subject. Since this material is covered extensively in the literature, the presentation will be cursory and glossary-like.

2.1 TERMINOLOGY

Structural optimization seeks optimal values of **design variables** which minimize or maximize a specific quantity termed **objective (or cost) function**, while satisfying a variety of behavioral and geometrical conditions, termed **constraints**. As such, modern structural optimization is, at least formally, in the realm of operations research, specifically mathematical programming.

The central engineering goal of designing efficient structures, has made structural weight the natural objective function to be minimized. This is especially true in the aerospace industry. Therefore, generally optimum design problems are dominantly minimum weight design problems, although other quantities such as reliability, deflection, stiffness, energy, and actual cost have been taken as objective functions as well.

Structural systems considered are usually idealized as discrete (beams, frames, grillages, arches, and rings) or continua (plates and shells).

Design variables may be cross-sectional variables or other parameters which describe the structural configuration and the material properties. Examples of cross-sectional variables are: cross-sectional areas (of truss members), section moduli (of frame members), and thickness (of plates). Examples of configuration parameters are topological variables (describing the connectivity among members and nodes) and shape (geometry) variables such as nodal coordinates.

A **constraint** is generally a restriction that must be satisfied for the design to be acceptable or feasible. Constraints may be of two types: **equality constraints** which are mostly behavioral constraints such as equilibrium, compatibility, and constitutive relations. Another example of equality constraints is a **linkage** constraint requiring identical design for several different members. **Inequality constraints** represent usually limitations imposed on the stresses, deflections, vibrational frequencies, buckling strengths, etc. A particular kind of inequality constraint is a **side constraint**, which imposes a direct limit on the range of variation of a design variable. A typical example is imposing a minimum and maximum value on cross-sectional areas or stresses.

2.2 THE MATHEMATICAL PROBLEM

In mathematical terms, structural optimization problems may be cast in the following general form:

$$\text{Minimize } W(\underset{\sim}{x}) \tag{1a}$$

$$\text{Subject to } g_j(\underset{\sim}{x}) \leq c_j \quad j = 1,\ldots,m \tag{1b}$$

$$\text{and } \underset{\sim}{x}^{(\ell)} \leq \underset{\sim}{x} \leq \underset{\sim}{x}^{(u)} \tag{1c}$$

where W is the objective function, $\underset{\sim}{x}$ is the vector of a n design variable $(x_1, x_2,\ldots,x_n)$, $g(\underset{\sim}{x})$ is a functional expression of the constraints, and c_j is the limiting value of the contraint. Equation 1c. includes the side constraints; the subscripts (ℓ) and (u) indicate lower and upper bounds respectively.

The n-dimensional space defined by $\underset{\sim}{x}$ is termed the **design space**. The constraints define hyper-surfaces in the design space. A set of design variables define a design point in the design space. Such a point is called a solution, which may be **feasible** if all the constraints are satisfied or **infeasible** if any of them is violated. Structural optimization problems usually give rise to **Nonlinear Programming** (NLP) problems. Because of this fact, it is generally not known whether the solution of the optimization problem constitute a **global** or **local** optimum. This mathematically disturbing fact, does not present severe difficulties in design practice, where even a relative optimum may offer substantial savings. Global solutions are found only for **convex** problems. A particular class of convex problems is the **Linear Programming** (LP) problems. Frequently nonlinear problems are solved as sequence of linear programming problems (SLP). Plastic design problems may also be cast as LP problems.

The solution of a nonlinear structural optimization problem typically involves an estimate of an initial solution, or design point, $\underset{\sim}{x}^{(0)}$ from which a search for optimum is started, using some gradient or steepest descent approach. The initial and intermediate design points do not have to be feasible. In many cases, it is convenient to incorporate the constraints of the problem and the objective function in a single function called Lagrangian and denoted by L, to be minimized.

$$\text{Minimize } L = F(x) + \underset{\sim}{\lambda} G(x) \tag{2}$$

where F (x) is the objective function of Equation 1a; the functions G (x) are the constraints of Equations 1b and 1c converted into equalities such that every constraint is satisfied if G (x) = 0; and $\underset{\sim}{\lambda}$ is a vector of Lagrangian Multipliers, with as many components as constraints. Thus, Equation 2 is equivalent to Equation 1, and, as an unconstrained problem, its solution may be formally obtained by solving the equations:

$$\frac{\partial L}{\partial x_i} = 0; \quad i = 1, 2,\ldots,n \tag{3a}$$

and

$$\frac{\partial L}{\partial \lambda_j} = 0; \quad j = 1, 2,\ldots,n \tag{3b}$$

---

*Numbers in parentheses indicate references given at the end of each chapter.

The formulation of Equations 2 and 3 is sometimes termed **classical formulation**. Equations 3 are termed the Kuhn-Tucker conditions (1)*. In practice, they may be used as necessary criteria for checking optimality or serve as a basis for calculating the optimum.

2.3 CLASSIFICATIONS

Structural optimization problems may be classified by the type of design variables, loadings, and constraints. In practice, each class usually requires a different method or strategy of solution. The following are examples of commonly encountered problems.

Shape optimization problems seek the optional **layout** of nodes and members in a discrete structure. Three types of variables are involved in such problems: (a) **geometrical** or **configurational** variables, such as nodal coordinates; (b) **topological** variables; these are 0-1 variables defining the connectivity of members to nodes; and (c) **sizing** variables for actual design of members. Because of the wide differences in the numerical behavior of these variables, their simultaneous inclusion in a single problem is often avoided. Fixed geometry is assumed in most of the problems recently being tackled.

Based on the loadings and constraints, problems are classified as **static** and **dynamic**. Static problems may include **stress** constraints as well as **displacement** and **buckling** constraints. Dynamic problems may be divided into problems with **response** or **frequency** constraints. Other categories are: optimization with **aeroelastic** requirements dealing with wings and other lifting surfaces; **reliability-based** optimization dealing with probabilities of failure as variables or as the objective function.

2.4 METHODS OF OPTIMIZATION

Due to the fact that no single procedure exists for solving all nonlinear programming problems (NLP), different methods and techniques are used depending on the problem at hand. The following is a brief listing of such widely used methods and techniques.

<u>Analytical and Numerical Methods.</u> The origins of modern optimization methods can be traced to the layout theory of Maxwell (1854) (2), Cilley (1900) (3), and Michell (1904) (4). This theory has been reconsidered recently by Prager (5), Cox (6), and Hemp (7), and developed in what is today termed **analytical methods**. These methods seek functions and functional relations, representing a structural design, such that weight is minimized. The results obtained are highly theoretical. In parallel to analytical methods, **numerical methods** began to develop over the last 20 years. These methods use mathematical programming techniques to optimize structures which can be modeled using finite element methods. The advent of the electronic computer has naturally caused these methods to develop very rapidly. It must, however, be emphasized that analytical methods are most important to the development and the evaluation of numerical methods.

Fully Stressed Design. This approach is widely used in design practice. It may be regarded as an extension of **simultaneous mode of failure** concept. This latter concept was widespread in the 1940's and 1950's and is represented in the works of Shanley (8), Gerard (9), and Cox (6), which also originated the current analytical methods. Whereas, in simultaneous modes of failure, a single loading condition is implied, in fully stress design failure in multiple loading conditions is desirable for optimality. Generally, the method is considered heuristic, although in some cases it can be proved to achieve optimum (10). The appeal of the method ies in its simplicity: the members are iteratively resized so that each member has maximum allowable stress in at least one loading condition. A discussion on the inadequacy of full-stressed design methods may be found in References 11 and 12.

Mathematical Programming. These methods were the chief vehicles for most of the recent developments in numerical optimization. Operations research general techniques, such as linear programming (LP) and non-linear programming (NLP), and specialized techniques, such as dynamic programming (DP), integer programming (IP), and geometric programming (GP)--were adopted and programmed on computers. Steepest descent or gradient methods may be also classified within mathematical programming. Examples are the gradient projection and feasible direction methods and a variety of sequential unconstrained minimization techniques (SUMT). The latter may also be regarded as an extension of the classical formulation since the objective function is augmented using products of a penalty function and the problem constraints.

A typical mathematical programming algorithm begins with an initial vector of design variables, $\underline{x}$. At iteration (n), $\underline{x}$ is updated by the relation

$$x^{(n+1)} = x^{(n)} = \alpha^* s^{(n)} = x^{(n)} + \Delta x^{(n)} \tag{4}$$

In this equation $s^{(n)}$ is the search direction vector and α^* is a scalar which determines the length of the move along that direction. $s^{(n)}$ is found such that for some small α, the design is improved while satisfying the constraints. Once $s^{(n)}$ is determined, a minimization problem with a single variable α is performed to determine α^*. The method of choosing the search direction and the interpolating scheme used to find the step size α^* distinguishes one mathematical programming method from another. One of the most significant features of these techniques is that no a priori assumptions are made regarding which constraints will be critical at optimum. A general review of these techniques may be found in Reference 12. A discussion on the development and use of mathematical programming may be found in References 14 to 16.

Optimality Criterion Method. These methods essentially involve the derivation of necessary and sufficient conditions for an optimum with specific design requirements and their association with a certain real or pseudo energy functional of the system. As such they may be viewed as a variation of the classical Lagrangian approach, requiring insight for selectively determining the Kuhn-Tucker conditions which are critical. The basic concept underlying optimality criterion methods may be found in References 17 to 22.

In the last 10 years, these methods gained wide acceptance and were extensively developed while methods of mathematical programming were consolidating. Appendix B is dedicated to recent developments of these methods (23-25) and to their relationship to mathematical programming methods.

Other Methods. Other methods have been adopted from other fields particularly from general optimization theory and analytical methods. In addition, techniques for decomposition and approximation were developed to simplify the solution and reduce cost. A special class of mathematical programming based problems utilized the concept of duality to improve solution efficiency. In most cases, when two or more methods are efficiently combined, a satisfactory solution may be achieved in as few as 10 iterations.

2.5 DEVELOPMENT AND LITERATURE

Schmit's work in 1960 (14) on structural synthesis and his subsequent pioneering efforts are agreed to be the beginning of modern developments of numerical structural optimization. Prior to this, in 1949 and 1956, mathematical programming was first applied to structural optimization problems by Prager (5) and Livesley (26). In 1964, Moses (27) introduced the technique of sequential linear programming (SLP). In 1968, Venkayya and his associates (28, 29) started the development of optimality criteria approaches, in which Prager (30) and Taylor (31) were instrumental. In approximately a decade (1960-1971) structural optimization developed into a full-fledged theory.

The first comprehensive survey in the field was written in 1963 by Wasiutynski and Brandt (32). Other notable periodic reviews and symposia were those by Sheu and Prager (33) in 1968, by Schmit in 1969 (15), by Pope and Schmit in 1971 (16), by Schmit in 1974 (34), by Reinschmidt in 1976 (35), and by Venkayya in 1978 (36).

The first books to present the methods and engineering applications of numerical optimization were written or edited by Cohn in 1969 (37), by Fox in 1970 (38), by Spunt in 1971 (39), and by Gallagher and Zienkiewicz in 1973 (40). Most recently, Kirsch (41) published a text book, particularly suitable for civil engineers, treating theory as well as applications. References 41, 40, and 9 are recommended as first books for the interested reader who is not familiar with structural optimization.

Chapter 8 includes comprehensive lists of books on general and structural optimization as well as a list of recent works covering the period between 1970 and 1980.

2.6 REFERENCES

1. Kuhn, H. W. and Tucker, A. W., "Nonlinear Programming," Proceedings of the Second Berkeley Symposium on Mathematical Statistics and Probability, (J. Neyman, ed.), pp. 481-493, University of California Press, 1951.

2. Maxwell, C., Scientific Papers II, 1869--Reprinted by Dover Publications, New York, 1952.

3. Cilley, F. H., "The Exact Design of Statically Indeterminate Frameworks, An Exposition of Its Possibility, But Futility," Trans. ASCE, Vol. 42, 1900.

4. Michell, A. G. M., "The Limits of Economy of Material in Frame Structures," Phil. Mag. Series 6, Vol. 8, 1904.

5. Greenberg, H. J. and Prager, W., "On Limit Design of Beams and Frames," Technical Report, Brown University, Office of Naval Research, Contract N7Onr-35806, Oct. 1949. Also, Prager, W., "Linear Programming and Structural Design," Rand Report P-1123, 1957.

6. Cox, H., The Design of Structures of Least Weight, Pergamon, Oxford, 1965.

7. Hemp. W. S., Optimum Structures, Clarendon, Oxford, 1973.

8. Shanley, F., Weight-Strength Analysis of Aircraft Structures, Dover, New York, 1960.

9. Gerard, G., Minimum Weight Analysis of Compression Structures, New York University Press, 1956.

10. Spillers, W. R., Interative Structural Design, North Holland, 1975.

11. Razani, R., "The Behavior of the Fully-Stressed Design of Structures and its Relationship to Minimum Weight Design," AIAA Journal, Vol. 3, No. 12, December 1965, pp. 2262-2268.

12. Kicher, T. P., "Optimum Design Versus Fully Stressed," Proc. ASCE Journal of Struct. Div., Vol. 92, No. ST6, 1966.

13. Vanderplaats, G. N., "Numerical Optimization - A Practical Design Tool," Proc. ASCE 7th Conference on Electronic Computation, St. Louis, Mo., August 1979.

14. Schmit, L. A., "Structural Design by Systematic Synthesis," Proc. of 2nd National Conf. on Electronic Computation, ASCE, 1960, pp. 105-132.

15. Schmit, L. A., Jr., "Structural Synthesis 1959-69: A Decade of Progress," published in: Recent Advances in Matrix Methods of Structural Analysis and Design, 1971.

16. Pope, G. G. and Schmit, L. A., editors, Structural Design Applications of Mathematical Programming Methods, AGARDograph No. 149, February 1971.

17. Venkayya, V. B., Khot, N. S., and Reddy, N. S., "Energy Distribution in an Optimum Structural Design," AFFDL-TR-68-156, March 1969.

18. Berke, L., "An Efficient Approach to the Minimum Weight Design of Deflection Limited Structures," AFFDL-TM-70-4-FDTR, May 1970.

19. Gellatly, R. A. and Berke, L., "Optimal Structural Design," AFFDL-TR-70-165, April 1970, Air Force Flight Dynamics Laboratory, Wright-Patterson Air Force Base, Ohio.

20. Kiusalaas, J., "Minimum Weight Design of Structures via Optimally Criteria," NASA TN D-7115, 1972.

21. Venkayya, V. B., Khot, N. S., and Berke, L., "Application of Optimality Criteria Approaches to Automated Design of Large Practical Structures," Second Symposium on Structural Optimization, AGARD-CP-123, Milan, Italy, April 1973.

22. Berke, L. and Khot, N. S., "Use of Optimality Criteria Methods for Large Scale Systems," ARARD Lecture Series No. 70, Structural Optimization, 1974.

23. Khot, N. S., Berke, L., and Venkayya, V. B., "Minimum Weight Design of Structures by the Optimality Criterion and Projection Method," AIAA 79-0720, Paper presented at AIAA/ASME/ASCE/AMS, 20th Structures, Structural Dynamics and Materials Conference, St. Louis, Mo., April 1979.

24. Fleury, C. and Sander, G., "Relationships between Optimality Criteria and Mathematical Programming in Structural Optimization," Proc. Symp. Applications of Computer Meth. in Engng. (Ed. C. Wellford, Jr.). University of Southern California, 507-520 (1977).

25. Fleury, C. and Schmit, L. A., "Prime and Dual Methods in Structural Optimization," Journal of the Structural Division, ASCE, Vol. 106, No. ST5, Proc. Paper 15431, May 1980, pp. 1117-1133.

26. Livesley, R. K., "The Automatic Design of Structural Frames," Quart. J. Mech. Appl. Math., 9, Part 3 (1956).

27. Moses, F., "Optimum Structural Design Using Mathematical Programming," J. of the Structural Division, ASCE, Vol. 90, No. ST6, December 1964, pp. 89-104.

28. Venkayya, V., "Design of Optimum Structures," Computers and Structures, 1, No. 1/2, 265-309 (1971).

29. Gellatly, R. A. and Berke, L., "Optimum Structural Design," AFFDL-TR-70-165, February 1971.

30. Prager, W. and Marcal, P., "Optimality Criteria in Structural Design," AFFDL-TR-70-166, May 1971.

31. Taylor, J., "Optimal Design of Structural Systems: An Energy Formulation," AIAA J., 7, pp. 1404-1406 (1969).

32. Wasiutyński, Z. and Brandt, A., "The Present State of Knowledge in the Field of Optimum Design of Structures," Applied Mechanics Review, Vol. 16, No. 5, May 1963, pp. 341-348.

33. Sheu, C. Y. and Prager, W., "Recent Developments in Optimal Structural Design," Applied Mechanics Reviews, Vol. 21, No. 10, October 1968, pp. 985-992.

34. Schmit, L. A. (editor), Structural Optimization Symposium, ASME Winter Annual Meeting, 1974 (AMD7).

35. Reinschmidt, K. F., "Review of Structural Optimization Techniques," Development of Computational Methods in Structural Analysis: Past, Present, and Future, Chapter V, Preprint 2765, ASCE Annual Convention, Philadelphia, 1976.

36. Venkayya, V., "Survey of Optimization Techniques in Structural Design", AFFDL-TM-FBR-78-43.

37. Cohn, M. Z. (editor), An Introduction to Structural Optimization, S. M. Study No. 1, University of Waterloo Press, Canada, 1969.

38. Fox, R. L., Optimization Methods for Engineering Design, Addison-Wesley, 1971.

39. Spunt, L., Optimum Structural Design, Prentice Hall, 1971.

40. Gallagher, R. H. and Zienkiewicz, O. C. (editors), Optimum Structural Design, Wiley, 1973.

41. Kirsch, U., Optimum Structural Design: Concepts, Methods and Applications, McGraw Hill, New York, 1981.

PART II
NUMERICAL METHODS

3. INDUSTRIAL APPLICATIONS

3.1 INTRODUCTION

In organizing examples of optimization of industrialized structures, some orderly division is necessary. It could be done on the basis of the mathematical programming methods used in the solution; but then many examples could equally be solved by several other methods. Furthermore, anyone contemplating optimizing a specific type of structure will probably find some unique features which require modifying existing or available mathematical programming algorithms. As a consequence, this chapter divides iterative structural decision problems into a hierarchy of three categories: component or element design, mass-produced structural systems, and specialized applications including discrete decision variables. It is recognized that these categories overlap especially in problems involving more complex conceptual and creative decision variables, such as nodal coordinates and material parameters, nevertheless these three categories do cover a very wide range of applications and will each be discussed below, citing actual examples or programs. It should be remarked that several applications cited here were selected from among many contributions of individuals, who responded to an announcement and correspondence soliciting such contributions.

3.2 COMPONENT DESIGN

Component design problems are characterized by a well-defined code of practice for the constraints and a relatively direct method of calculating design loads such as moment, shear, torsion, and thrust. Some reported examples of application include welded wide flange beams, unsymmetrical box girders and prestressed tee-beams (1-8). Design variables are typically depth, thickness, shape, reinforcement ratio and other design details which are often part of the tedious process of redesign, and for which economic selection rules are not always available. These elements typically have a small number of independent design parameters (less than 10) but relatively complex functional code requirements such as allowable buckling and displacement constraints. As an example, the American Institute of Steel Construction (AISC) code specifies lateral buckling constraints which are discontinuous, representing a transition from elastic to inelastic behavior (1).

In some applications it becomes necessary to repeat the design for a large number of different elements. For example, a particular overhead crane manufacturer using box girder sections needed over 5,000 specified designs (2). This obviously required an efficient design program.

Several computer packages are available to automate element optimization. One approach uses the penalty technique which combines the criteria function (cost or weight) and the constraints into a single expression to be minimized (9). This transforms the more difficult nonlinear programming problem with cost and constraints into a more tractable unconstrained minimization for which many straightforward solution methods are available. Examples of other successful approaches are linear approximation and geometric programming (10).

In mathematical form the component design problem has the same general format presented in section 2.2. In practical applications the optimum values of the variables must be rounded-off to the nearest available sizes usually by a search in the neighborhood of the absolute minimum. Common characteristic of such component design problems is that the user may have a variety of different cases to investigate. Thus the choices are an interactive-type program in which the designer can input specific requirements such as span, load, material properties, or produce a complete catalog of possible design to cover the spectrum of applications. The latter is often a valuable tool for estimating or sales purposes. Use of component optimization will be illustrated with several examples. It is hoped that these examples will encourage the use of structural optimization.

3.2.1 Box-Girder Design

Structures designed by civil engineers are often one-of-a-kind items so that the specialized programming efforts justified for a specific structure are limited. However, overhead cranes using welded box-girder cross-section are one example where the structure is an "off the shelf item," with different solutions corresponding to span, capacity, trolley configuration, etc. Such designs have been performed for fabricators in Ohio, California, Canada, and elsewhere, requiring over 20,000 separate results so that the optimization programs had to be extremely efficient to keep design costs within reasonable limits. The result was a catalog of standard designs for marketing use.

A typical cross-section of crane box-girder cross-section has five design variables: web depth (H), flange width (B), equal web thickness (t_w), top flange thickness (t_t), and bottom flange thickness (t_b). The girder itself is, however, statically determinate with constraints on stress, displacement and flange buckling precisely spelled out in the applicable design code (11). In most cases, the optimum design is unsymmetric and therefore normally presents a difficulty for the designer to determine the best design by hand calculation. The objective function to be minimized is weight (W) which is proportional to $[2Ht_w + B(t_t + t_c)]$.

Other crane examples have included stiffened webs and flanges, unequal webs with unsymmetrical rail placement, and flanges with channels for reinforcement.

3.2.2 Welded Plate Girder

In this case it is sought to automate the selection of the minimum-weight cross-section for a welded wide-flange plate girder, designed according to the AISC specification (12), including lateral buckling constraints. There are four design variables: web depth and thickness, and flange width and thickness. For laterally unbraced beams the number of design variables cannot be reduced by considering flange area as a single variable.

As an example, in one particular project the project was done in about 10 days by an engineering student with little knowledge of mathematical programming. The principal difficulty encountered was in the interpretation of the AISC constraints and recognition that there are implied discontinuities in the allowable stress (1). One aspect of this design example is the difficulty of a direct design procedure to specify the active constraints and the use of these relationships to proportion the flange and web. Even in the absence of lateral buckling such an approach was seen to be a tedious combination and elimination of constraints (13). Available solutions are often in the form of tables and graphs which become obsolete with changes in the specification of material properties in the code.

The optimization approach permits the user to simultaneously include all the conceivable constraints and have the computer program find the optimum design and the active limitations for a given loading and span. Changes in code formulae merely require changes in constraint expressions. The program described above was subsequently extended to include the design of welded wide-flange sections using hybrid steel with cost as the objective function to be minimized.

3.2.3 Standardized Frames

The welded plate-girder design program previously described was used together with a column design program in a larger program for automatic design of welded gabled frames (4). The latter has practical applications since many manufacturers and fabricators maintain design tables of such frames for different geometry and load configurations for use in mill buildings and warehouses. In a symmetrical frame with prismatic members there are eight design variables for girders and columns. The penalty method can still be used despite its being a statically indeterminate frame. Formulas for gabled frames are available to express reaction forces and moments in terms of member stiffnesses without solving simultaneous equations. Otherwise an analysis-design iteration may be used since reaction forces are relatively insensitive to changes in member sizes. Usually only several force analyses are needed. Load conditions include gravity, snow, dead load (crane), wind, and live load. Parametric studies may be performed to show the effect on weight or cost of such parameters: frame spacing, member depth limitation, deflection constraint, and roof pitch. In this manner, the automated optimization process can be extended to include a broader group of design parameters. Recently, extensions of gabled frame optimization have been reported for nonprismatic tapered members including appropriate provisions for checking their design (14, 15).

3.2.4 Reinforced Concrete Beams

In any building design programs, rules are specified to design the component details. For example, one reinforced concrete building design program uses, for reinforced concrete beams, one-half the balance steel ratio and a width to depth ratio of 0.5. With little effort, however, a program was written to provide the optimum dimensions and steel area. By examining a range of variables the program could also determine optimum concrete strength and mix design

(8). In fact, a study of the optimum design range should lead to establishing more accurate design rules which previously have been rather arbitrary.

3.2.5 Shear Wall Design

The design problem in this case is typically the proportioning of the shear wall cross-section to resist lateral and vertical loads. As an example, in an actual program the penalty method determines cross-sectional depths, widths, and steel ratios. The wall is designed to resist moment in one direction. The cost function includes concrete, steel, and forming costs. This is another example where the load analysis is realtively simple and needs to be done only a small number of times. The program is especially promising in cases where the architect has a number of possible locations for shear walls and the program then determines their optimal locations and sizes (6).

3.2.6 Prestressed Concrete Beams

Eleven design variables characterize this example including cross-section geometry, tendon layout, and prestress force. Goble has shown how the minimum cost beam can be found including constraints on stress, deflections, and moments following standard code practice (3).

3.2.7 Connection Design

Optimum connection design is closely related to component design to the sense that the number of variables is relatively small and the constraints are numerous and specified by some code or specification. In one reported example (16), bolted and welded frame connections are designed to AISC specifications using an interactive computer program. The method used for optimization is linear programming on a linear approximation of the problem.

3.3 MASS-PRODUCED STRUCTURAL SYSTEMS

Component design is characterized above by having known input forces and a need to proportion the element to carry these forces. Since the generally available optimization programs usually require an exhaustive search for the optimum by checking a large number of possible designs the procedures are not useful when the forces must be calculated for each design check from a matrix structural analysis. Other optimization procedures must also be utilized when the structure contains several components, which all interact with structural response constraints such as stability, deflections, vibration, or dynamic response. In addition, such "system" optimization arises when some of the design variables affect more than one element such as geometry, material, or topology. In these cases, the force distribution requires a matrix or finite element analysis for each cycle of design check. Computer costs usually restrict the number of possible trials that the computer can sequentially examine as it searches for the optimum to often under 10 trials. Some cases that have been optimized include foundation or ship grillages, trusses and frames (17-19). Search techniques for systems often utilize linear programming or heuristic searches and hence, large changes in design may be made with only a single analysis iteration (18, 19).

In some cases, the design space can be reduced to only several variables by creating an iteration between system and component design. That is, for a given set of system variables the cost (or weight) is computed by optimizing each element independently. For example, a roof truss with depth and number of panels being the system variables and component cross-sections the different element variables. Applications such as these are aided by having element "libraries" to provide optimum elements over the possible range of input variables (20). This subdivision of system and component design spaces is more difficult to generalize, however, when the constraints involve both sets of variables simultaneously. The following are some examples.

3.3.1 Grillages

Orthogonal beam systems are used in ships, foundations, and other applications to carry load. The behavior interaction of elements precludes their being optimized by fully stressed procedures since convergence to even a good design cannot be guaranteed. Penalty and linear programming methods have been successfully used on these examples (17).

3.3.2 Transmission Towers

An excellent example of industrialized construction is a transmission tower which is usually produced in large quantity in several basic designs. In addition to member sizing, optimization has been used to find geometric layout usually with minimum weight criteria. The simplest scheme is to separate geometric design variables and iterate alternatingly between a geometric design space and a member design space. Minimization in the former can be done by direct search or gradient directions (19, 21). Unpublished reports exist on procedures of optimum design of tapered members with special applications to tapered transmission poles, where both stress and load buckling constraints are considered. Guyed poles with nonlinear, large deflections have also been reported. Curve fittings of a shape profile effectively reduces the potentially large number of design variables in the cases.

3.3.3 Roof Trusses

Planar and three-dimensional roof layouts have been studied. In most cases, element sizes with fixed geometries have been found to satisfy both element constraints as well as system stiffness. The latter can be important for long-span roofs. In addition, geometrical variables such as overall depth and panel layouts have been included in optimization. If equal panel spacing is used then a planar truss only has two geometrical variables lending itself nicely to separating geometrical and element design as in the preceding example (20).

Thomas and Brown reported on a extensive study of roof optimization with cost criteria, realistic code constraints, and accurate member modeling especially for effective length values (22). The truss roof optimization considered available sections and joist variables but with fixed geometry. Lipson has also studied a variety of truss roofs looking at geometry, weight minimization, and cost optimization (23). Using a heuristic design space search known as the complex method member stresses and displacements are controlled and member sizes and coordinates found.

3.3.4 Prestressed Concrete Bridge Decks

A variety of configurations using solid sections or box sections have been designed by Bond using the SUMT approach (24). Cost criteria along with slab thickness, web thickness, reinforcing and prestressing steel design variables have been considered with constraints based on British Standards. Recently, transverse effects, cross-sectional distortions, shear lag, and torsion have been added to the program. Geometric programming has been presented by Ramamurthy for minimum cost design of prestressed concrete slabs (25). By using a transformation, Kirsch used a linear programming approach to optimize tendon configuration and prestress force in prestressed concrete systems (26). Included are non-uniform cross-sections, multiple load conditions, and variable prestressing forces. Geometric programming has also been used by Templeman to optimize concrete cellular spine beam decks (27). Design variables include the number of cells, top and bottom flange thickness and web thickness for each cell, transverse, reinforcement, longitudinal prestress area and forces, haunch lengths, etc. Final detail optimization is not the goal but rather an attractive design that can be modified or rounded-off by the designer. Codes of practice form the constraints while total cost is the objective function. Such bridges are used in large quantities in Great Britain and elsewhere. A new form of popular construction for builders is sequential box girders. Miller has reported on optimization of substructures especially the complex staging aspects of the constraints (28).

3.4 SPECIALIZED APPLICATIONS

This category of optimization is considerably more complex than either of the first two categories. For example, discrete design variables are not always clearly defined or constant in number and the cost function may be extremely complex and discontinuous. The discrete nature of variables often requires heuristic or intuitive search methods of solution which make general purpose programs irrelevant. One method, however, which has solved a variety of such problems is the dynamic programming technique which is easy to program if the problem formulation meets its basic definition (29). Among the reported structural design problems solved by dynamic programming are the minimum cost of continuous coverplated highway bridge girders, single-story building selection of different roof, column and foundation elements, spacing of supports of multispan girders, thickness variations in ship plate components, reinforcing bar arrangement in continuous reinforced concrete beams and girder selection for minimum material, detailing, and fabrication costs (20-24). These examples have in common discrete variable selection and more important a sequencing of decisions into stages which satisfy the dynamic programming criterion.

3.4.1 Design of Continuous Highway Bridge Girders

In the construction of highway bridges, including many interchange bridges, the continuous welded plate girder is a solution that is frequently selected. In this case the problem cannot be stated within the nonlinear programming framework because of its discrete nature. However, due to the number of elements of this type which are built, a much larger investment in computer programming is justified. The analysis of such element is in itself a rather complex task, in that the girder must be designed for an envelope of shear and moment forces produced by a number of different load conditions resulting from sets of moving loads.

The basic dynamic programming algorithm for this problem has already been stated in Reference 30. The design variables in the problem consists of the girder depth, web thickness and flange width and a series of flange thicknesses. One interesting aspect is that the decision must be made to use a uniform flange thickness or to change the thickness at locations which must be selected along the flange. This can be accomplished easily by groove welding different size plates together. If flange splices are used the minimum weight design is a design in which the flange changes thickness continuously along the span. This would produce a design that might be described as a fully stressed design in that the flange would be stressed to its allowable limits in every location in at least one load condition. Of course, such a design would not be practical to build and instead the flange thickness must change at discrete locations.

The problem is solved using an iterative approach, various portions of which contain elements of optimization techniques. Minimum cost is used as the criterion.

This same basic algorithm has been expanded to include the design of welded hybrid plate girders, plate girders of variable depth, and coverplated rolled beams (30, 31). Various versions of these programs have been used for a number of years by the Ohio Department of Transportation and other organizations in the routine design of girders. In this example program, a very large and complex code is produced which requires a substantial effort to prepare and maintain. It would certainly not be justifiable to use such a program for a structure that does not require substantial design effort using traditional means.

A heuristic approach was developed for finding optimal plate girders for railroad bridges by Rice. Girder depth and flange areas are varied to control stress and deflection. Some 288 superstructures of various spans and grades of steel were studied in this manner (32).

3.5 SHAPE OPTIMIZATION

The potential savings offered by shape optimization are generally more significant than those resulting from fixed-geometry optimization.

Preliminary design is the main area where the optimum configuration concepts should be applied. Its development generally originates in layout theory of analytical methods discussed in Chapter 9. This section concentrates on several aspects of shape optimization because the increasing needs for applications in this area.

3.5.1 Difficulties

The fact that shape optimization did not enjoy the same degree of progress and application as fixed-geometry optimization may be attributed chiefly to the following factors: (a) the increased number of decision variables, (b) the different degree of nonlinearity in the numerical behavior of these variables, and (c) the poential change in topology under both fixed and variable geometry optimization. These difficulties prevent the effective utilization of currently available geometric optimization algorithms in design practice.

3.5.2 Topological Considerations

Optimizing the topology, or the connectivity of a skeletal structure has been and still is a major stumbling block of structural optimization. While methods for mixed-integer-programming have been developed, the subject is still too complex for practical engineering applications. Two widely used methods of topological optimization are linear programming (33-35) and iterative stress-ratio or fully stressed design (35). In these methods, a set of redundants is typically eliminated from a larger admissible set of bars connecting all admissible nodes. Most known geometric optimization programs, however, assume the topology of the structure to be fixed; some do not even allow for vanishing cross-sections. Such programs may be quite useful for optimizing a known topology. In preliminary design, the configuration problem (geometry and topology) is critical.

3.5.3 Methods

Assuming the topology to be known, the nonlinear programming formulation of geometric optimization can be conceptually cast in the following form:

$$\text{Minimize } W(A, Q) = \sum \rho L_i A_i \quad \text{Subject to } g(A, Q) \geq 0$$

where W is the weight of the truss; A and L are the vectors of member cross-sectional areas and lengths, respectively; Q is the vector of nodal coordinates and ρ is the density of the material. The set of constraint functions, g, are behavioral constraints, such as equilibrium, compatibility and constitutive relations and side constraints restricting stresses and displacements as well as A and Q to vary between specified upper and lower bounds.

To avoid some of the difficulties cited above, it is not advantageous in common practice to optimize all variables simultaneously. This is commonly done by performing the optimization in two separate spaces: the sizing space, where A is optimized, assuming the geometry, or Q, to be fixed; and the coordinate space, where Q is allowed to vary. The process in all these techniques is iterative and continuously alternating between the A and Q subspaces until convergence is reached. The sequence in which the optimization is carried out in the two subspaces differs depending on the techniques used.

It is important to note that while member sizes may be optimized under either fixed or variable geometry, the optimization of geometry must always be accompanied or followed by member proportioning. This necessity is due to changes in the equilibrium equations and member forces resulting from the change in coordinates. Variation of node coordinates is usually based on a search along a certain direction, indicated by the particular gradient approach used. Since several steps, ΔQ, are usually taken along any such direction, a corresponding number of resizing and reanalyses must be performed, requiring tens or hundreds of analyses, which may be quite costly in a sizable problem.

A representative method of geometric optimization is the technique used by Vanderplaats (36). A special-purpose program for trusses, SADT (37), calls a general-purpose nonlinear optimization program CONMIN (38) whenever the coordinates are modified; thereupon the member proportions are updated and the truss is reanalyzed. Size optimization is performed under fixed geometry using the stress ratio method or a gradient method on the cross-section reciprocals.

A variation of this method is the process used by Spillers (39-40), where the geometry is varied in steps ΔQ obtained from a set of linearized Kuhn-Tucker optimality criteria. Hence, no search along a direction is involved, and the number of analyses is substantially reduced. Instead, a trial-and-error process must be performed to find the correct, Q whenever the standard step proves to be too large. Lipson's adaptation of the "complex" method (41) to optimum geometry problems, is another example of a technique which avoids gradient or directional search. An investigation of these processes has been discussed by Lev (42).

Most recently, a new technique was developed by Imai and Schmit (43). The method is based on the general multiplier method of optimization theory which is adapted to simultaneously handle design as well as geometry variables.

3.5.4 Applications

Shape optimization techniques are still in the stage of early development and their implementation in design practice is relatively rare, even though its potential applications are numerous. The following is a list of recent application and research works.

Control of stresses by tailoring a structural shape to an application has received attention for dams, pressure vessels, and other structures. Vitiello has illustrated the problem with dams (44). Maier studied plane strain and plane stress components modeled as rigid plastic structures (45). Middleton has looked at stress concentration factors and pressure vessels (46). Bond has concentrated on shape optimization for concrete elements and has looked at arches, stairways, shell roofs, water tanks, and arch dams (47). Somekh and Kirsch (48) utilize various methods of optimization and computer graphics to solve problems involving minimum weight trusses with optimal topology and geometry. These results are not yet at the stage for direct implementation but show great promise for future development.

3.6 REFERENCES

1. Goble, G. G. and Moses, F., "Automated Optimum Design of Unstiffened Girder Cross Sections," AISC Engineering Journal, April 1971.

2. Goble, G. G. and Moses, F., "Experience with Practical Applications of Structural Optimization," Proc. 6th Conf. on Electronic Computation, ASCE, 1974.

3. Goble, G. G. and LaPay, W. S., "Optimum Design of Prestressed Beams," J. Am. Concrete Inst., Vol. 68, No. 9, September 1971.

4. Joffe, I., "Minimum Weight Design of Welded Gabled Frame Structures," Report No. 51, Dept. of Solid Mechanics, Struct. and Mech. Des., Case Western Reserve University, January 1972.

5. Moe, J., "Design of Ship Structures by Means on Nonlinear Programming Techniques," Symp. on Struct. Opt., AGARD Conf., Proceedings No. 36, 1969.

6. Stoman, Sayed, "Optimization of Shear Wall Structures," Report. No. 58, Dept. of Solid Mechanics, Struct. and Mech. Des., Case Western Reserve University, Cleveland, Ohio, June 1974.

7. Kirsch, U., "Optimum Design of Prestressed Plates," Journ. of the Struct. Div., ASCE, Vol. 99, ST6, June 1973.

8. Moses, F., "Optimization of Reinforced Concrete and Other Structural Elements," Symposium on Optimization and Automated Design of Structures. Report No. SK/M21, Div. of Ship Structures, Tech. U. of Norway, Trondheim, January 1972, pp. 281-297.

9. Fox, R. L. Optimization Methods for Engineering Design--Addison Wesley, 1971.

10. Templeman, A. B. and Winterbottom, S. C., "Structural Design Applications of Geometric Programming," Second Symposium on Structural Optimization, Milan AGARD CP No. 123, April 1973.

11. Specification for the Design, Fabrication and Erection of Structural Steel for Buildings, 7th ed., American Institute of Steel Construction, New York, New York, 1969.

12. "Specification for Electric Overhead Traveling Crane," CMAA Specification No. 70, Crane Manufacturers Association of America, Inc.

13. Holt, E. C. and Heithecker, G. C., "Minimum Weight Proportions for Steel Girders," Journal of the Structural Division, ASCE, Vol. 95, No. ST10, Proc. Paper 6838, October 1969, pp. 2205-2217.

14. Miller, C. J. and Moll, T. G., Jr., "Design of Tapered Member Gabled Frames," Proceedings of the Structural Stability Council, 1978.

15. Miller, C. J. and Moll, T. G., Jr., "Automatic Design of Gabled Frames," Proceedings, ASCE Seventh Conference on Electronic Computations, St. Louis, August 1979, pp. 134-149.

16. Douty, R., "Design of Steel Connections by Math. Programming," Journal of the Structural Division, ASCE, Vol. 106, No. ST5, Proc. Paper 15397, May 1980, pp. 1135-1154.

17. Moses, F. and Onada, S., "Minimum Weight Design of Structures with Application to Elastic Gillages," Intl. Journ. for Num. Methods in Eng., Vol. 1, 311-331, 1969.

18. Reinschmidt, K. F., Cornell, C. A., and Brotchie, J. F., "Iterative Design and Structural Optimization," Journal of Struct. Div., ASCE, Vol. 92, ST6, December 1969.

20. Moses, F., Miller, C., and Yeung, J., "Optimum Design of Building Structures," Computers in Structural Engineering Practice, Proceedings of the CSCE Specialty Conference, Montreal, October 1977.

21. Vanderplaats, G. N. and Moses, F., "Automated Optimal Geometry Design of Structures," Journal of the Struct. Div., ASCE, Vol. 98, ST3, March 1977.

22. Thomas, Jr., H. R. and Brown, D. M., "Minimum Cost Design of Truss Roof Systems by Nonlinear Programming," National Structural Engineering Conference, an ASCE Struct. Div. Specialty Conference, presented 22-25 August 1976, Madison, Wisconsin.

23. Lipson, S. L. and Russell, A. D., "Cost Optimization of Structural Roof Systems," Journal of the Struct. Div., ASCE, Vol. 97, No. ST8, August 1971, pp. 2057.

24. Bond, D., "An Examination of the Automated Design of Prestressed Concrete Bridge Decks by Computer," Proceedings Institution Civil Engineers, December 1975, Vol. 59, pp. 669-697.

25. Ramamurthy, S., "Optimum Design of Prestressed Concrete Slabs Using Primal Geometric Programming," International Journal for Numerical Methods in Engineering, Vol. 13, December 1978, pp. 229-246.

26. Kirsch, U. "Optimized Prestressing by Linear Programming," Int. Journ. for Num. Methods in Engineering, Vol. 7, 1973, pp. 125-136.

27. Templeman, A. B. and Winterbottom, S. K., "Optimum Design of Concrete Cellular Spine Beam Bridge Decks," Proceedings, Institution of Civil Engineers, Longon, Part II, June 1979.

28. Miller, C. J. and Accicly, "Preliminary Design of Precast, Segmented Box-Girder Using Optimization," *Proceedings, Computer in Structural Engineering Practice*, CSCE, Montreal, October 1977.

29. Bellman, R. and Dreyfus, S. E., *Applied Dynamic Programming*, Princeton Univ. Press, Princeton, J.J., 1962.

30. Goble, G. G. and DeSantis, P. V., "Optimum Design of Mixed Steel Composite Girders," *Journal of Structural Division, ASCE*, Vol. ST6, December 1966.

31. Hsu, K.-y, "Optimum Design of Highway Bridges," thesis presented to Case Western Reserve University, at Cleveland, Ohio in 1974, in partial fulfillment of the requirements for the degree of Doctor of Philosophy.

32. Rice, J. C., "Optimum Design of Continuous Plate Girders for Railroad Bridges," Masters Thesis presented to the University of Virginia, May 1976.

33. Dorn, W. S., Gomory, R. E., and Greenberg, H. J., "Automatic Design of Optimal Structures," *Journal de Mechanique*, Vol. 3, 1964, pp. 25-52.

34. Moses, F., "Optimum Structural Design Using Linear Programming," *Journal of the Structural Division, ASCE*, Vol. 90, No. ST6, December 1964, pp. 89-104.

35. Spillers, W. R., *Interative Structural Design*, North Holland Publishing Co., 1975.

36. Vanderplaats, G. N., "Design of Structures for Optimum Geometry," NASA Technical Memorandum TM X-62,462, August 1975.

37. Vanderplaats, G. N., "SADT" (Structural Analysis and Design of Trusses), a computer program developed at NASA Ames Research Center, March 1979.

38. Vanderplaats, G. N., "CONMIN - A FORTRAN Program for Constrained Function Minimization, User Manual," NASA Technical Memorandum TM X-62,282, August 1973.

39. Spillers, W. R., "Iterative Design for Optimal Geometry," *Journal of Structural Division, ASCE*, Vol. 101, No. ST7, July 1975, pp. 1435-1442.

40. Spillers, W. R. and Kountouris, G. E., "Geometric Optimization Using Simple Code Representation," *Proc. Journal of the Structural Division, ASCE*, May 1980.

41. Lipson, S. L. and Gwin, L. B., "The Complex Method Applied to Optimal Truss Configuration," *Computers and Structures*, 7, 3, 461-468, June 1977.

42. Lev, O. E., "Sequential Optimization of Structural Geometry," Merritt CASES, Inc., Technical Report to NSF Grant. No. PFR 78-04313, April 1980.

43. Imai, K., "Configuration Optimization of Trusses by the Multiplier Method," Report No. UCLA-ENG-7842, Mechanics and Structures Department, School of Engineering and Applied Science, University of California, Los Angeles.

44. Vitiello, E., "Shape Optimization Using Mathematical Programming and Modeling Techniques," Second Symposium on Structural Optimization, AGARD Conference Proceedings, No. 123, Milan, Italy, April 1973.

45. Maier, G., "Limit Design in the Absence of a Given Layout: A Finite Element, Zero-One Programming Approach," Journal of Structural Mechanics, 1(2), pp. 213-230, 1972.

46. Middleton, J. and Owen, D. R. J., "Automated Design Optimization to Minimize Shearing Stress in Axisymmetric Pressure Vessels," Nuclear Engineering and Design, 44, 1977, pp. 357-366.

47. Bond, D., "Computer Aided Design of Concrete Structures Using Isoparametric Finite Elements and Nonlinear Optimization," Proceedings Institution Civil Engineers, Part 2, Vol. 67, September 1979.

48. Somekh, E. and Kirsch, U., "Structural Design Using Interactive Optimization," Proceedings, Seventh Conference on Electronic Computation, St. Louis, August 1979, pp. 168-189.

4.0 AEROSPACE APPLICATIONS

4.1 HISTORICAL OVERVIEW

Throughout the aerospace industry, whether in aircraft, space or missile design, a dominant theme in design has always been minimum weight. This is a natural result of the fact that a 1% reduction in structural weight typically results in a 3% reduction in vehicle gross weight or a 2% increase in payload for the same gross weight. Therefore, it is natural that the aerospace industry has played a major role in the development of structural synthesis techniques.

Historically, however, the early work in structural synthesis was applied to nonaerospace structures where plastic design techniques were applicable (see for example, references 1-3). This was a logical beginning because, as it turned out subsequently, plastic design may be cast as a linear programming problem, for which the general solution techniques were available.

Although not the first application of nonlinear numerical optimization methods to structural design, Schmit's work in 1960 (4) began an era of intense research in structural synthesis. Shortly after this, applications to aerospace structures began in earnest, motivated in large part by the space race. This effort resulted in numerous applications of numerical optimization techniques to stiffened plate and shell structures for minimum weight (5-10). The resulting design programs were usually special purpose programs developed to solve a specific design task. By the late 1960's, emphasis had turned from spacecraft to aircraft applications, initially with special purpose programs (11-16) and later with emphasis on the use of general finite element analysis capabilities (17-23).

The general trend in the 1960's was toward the solution of increasingly complex design problems in the sense of dealing with multiple loading conditions, post-buckling behavior, dynamics, and aeroelastic constraints. In most cases the number of independent design variables was relatively small, typically fewer than 20. As interest in automated design grew, emphasis moved from special purpose programs to general application programs based on finite element analysis. However, as larger problems were attempted, it became apparent that the cost of automated design using numerical optimization grew exponentially with the number of independent design variables (24).

In response to this apparent impasse to further development of the state-of-the-art, the optimality criterion approach (25-28) gained popularity and dominated the literature of the early 1970's. While these methods seemed to lack some of the generality of the nonlinear programming methods, they were able to design significantly larger structures. The essential difference between nonlinear programming and optimality criterion is that nonlinear programming makes no assumptions about which or how many constraints are binding at the optimum while optimality criterion define the conditions to be satisfied at the optimum and develop a recursive design procedure based on that set of conditions.

The mid-1970's was a time of consolidation in which optimality criterion methods were applied to an increasingly wide class of problems, including design of structures subject to aeroelastic constraints and structures made of composite materials. During this time other researchers were pursuing ways to design more practical structures within the nonlinear programming framework. These efforts resulted in the now familiar approximation concepts which can be used to greatly reduce the number of detailed analyses required to reach the optimum design (29).

The last few years have seen increased understanding of the design philosophy to be used in an automated design environment and the similarities and differences in nonlinear programming and optimality criterion methods are becoming better understood (30-33).

4.2 APPLICATION LITERATURE REVIEW

The degree of specialization which is associated with aerospace applications depends on the type of constraints which are imposed on the design problem. There are constraints such as displacement and stress constraints which are typical also of general nonaerospace problems. There are also constraints such as static and dynamic aeroelastic constraints which are almost exclusively encountered in aerospace applications. Here it is often found that special methods, tailored to such constraints are used instead of more general methods. For this reason, the following discussion of aerospace applications is divided into sections according to the type of constraints.

Further distinction made in the following discussion concerns the closeness of the application to actual flight structures. Papers which present methods and algorithms and use as examples highly simplified structural models are quite common. Only few papers, however, describe applications to structural models which are close in detail and complexity to the models used for the analysis of such structures. Papers describing applications of optimal design methodology to the actual design of an aerospace structure are very rare. In the following, these last two categories of applications are emphasized.

The methods developed for the optimization of structures with stress, displacement, and system stability constraints are applicable to aerospace structures as well as nonaerospace structures. From the previous survey papers and the most recent survey paper, it is evident that the number of publications in the field of structural optimization is very large. This section includes only references containing direct applications to aerospace structures. For completeness of applications, earlier papers presenting concepts which are still applicable in the present state-of-the-art were not excluded.

4.2.1 Stress and Displacement Constraints

For the minimum weight design of structures subject to stress constraints, it seems that the fully stressed design (FSD) approach is the rule (see for example, Lansing and coworkers, References 34 and 35). The alternative of using mathematical programming methods becomes prohibitively expensive when hundreds or thousands of individual elements are to be sized. In using FSD for aerospace structures we face the additional problem that the loads depend on the deformed shape of the structures. Giles and coworkers (18, 36) developed strategies for interfacing the FSD iterative process with the iterative calculation of loads and deformed shapes. The procedure was applied for a design study conducted at NASA Langley Research Center of a fairly realistic model of a supersonic cruise aircraft which was modeled with 746 grid points and 2,369 elements of which 718 were resized.

A variant of FSD for structures subject to high thermal stresses called thermal fully stressed design (TFSD) was developed by Adelman and coworkers (37, 38). However, recent experience with both FSD and TFSD showed that the method may lead to bad designs for highly redundant structures, especially when composite materials are used.

An early nonlinear mathematical programming type method developed by Gellatly, Gallagher, and Luberacki (Reference 39) and Gellatly (Reference 40) consists of three modes of travel: (1) an initial step, (2) steepest descent step, and (3) a side step. Initially the FSD step is used to select a better first guess design. The main procedure then successively uses the steepest descent move to travel towards the constraint surface and a side move to avoid a constraint violation. In Reference 57 a metal multispar wing structure is designed with stress and displacement constraints.

Taig and Kerr (41), Venkayya and coworkers (42), Isakson and coworkers (23), and Sander and Fleury (43) present applications of the FSD plus optimality criteria methods to wing or tail structures. One example of Isakson (23) is a model of a bomber fin that has 375 nodes and 1,293 members. The model was based on an early design model of an actual structure.

Melosh and Luik in Reference 44 use the intuitive optimality criterion type approach based on the principle that stresses in a member are usually primarily affected by the member's size. Each element of the structure is designed in sequence until the minimum weight design is reached. The time consumed in the reanalysis of the structure after each element resizing is reduced by a rapid reanalysis procedure. Johnson, Melosh, and Luik (Reference 45) have used this procedure to design an idealized swept wing.

Dwyer, Rosenbaum, Shulamn, and Pardo in Reference 46 present the FSD procedure as applied to airframe redundant structures. Two algorithms based on "average stress" and "nodal stresses" are described. The "nodal stresses" approach is found to give a smoother distribution of the material than the 'average stress' method. Application of the two approaches is illustrated by the design of a metal wing structure.

Dwyer, Emerton, and Ojalvo in Reference 47 develop a procedure for the design of aircraft structures with stress and displacement constraints. The algorithm is based on the modified FSD technique for the stress constraint problem. The displacement constraint problem is solved by using the gradient procedure discussed in Reference 25. A computer program ASOP-1 incorporating these algorithms was written, and its capability was illustrated by designing a wing structure and horizontal stabilizer. The horizontal stabilizer was idealized with 890 elements and had 1,171 degrees of freedom. In ASOP-1 plate elements subjected to a compression load can be designed by using the stability tables.

Lansing, Dwyer, Emerton, and Ranalli in Reference 48 discuss the application of the FSD procedure for designing a wing stabilizer with a metal and fiberous composite skin. The number of lamina in the 0°, 90°, and $\pm 45^o$ direction for each is determined by using the forces N_x, N_y, N_{xy} acting on each element. The optimum distribution of lamina in the four fiber directions for each element is determined in turn by incrementally changing the number of lamina in each fiber direction and selecting the one that gives the minimum weight for that element and satisfies the strength criteria.

Khot, Venkayya, Johnson, and Tischler in References 49 and 50 present a method based on the strain energy distribution and a numerical search in the design of minimum weight metal and fiber reinforced composite structures. The stress constraint problem is solved by using the recurrence relation based on the optimality criterion. The optimality criterion used can be stated as "The optimum structure is the one in which the average strain energy density is the same for all layers and for all elements." The concept of designing a structure based on the constant strain energy for a truss was first proposed in Reference 11. In Reference 50, a metal and composite wing structure idealized with 170 members and 264 degrees of freedom is designed to satisfy the stress and displacement constraints. The displacement constraints are satisfied by using the gradient method discussed in Reference 25.

Giles, Blackburn, and Dixon in Reference 18 have developed a system of computer programs to automate the preliminary structural design of a complete aerospace vehicle. Analysis is based on using NASTRAN (level 12) and sizing of the elements is based on the FSD technique. Results are presented for the preliminary design study of a Mach 3 transport wing.

Sobieszczanski and Loendorf in Reference 19, and Sobieszczanski in Reference 51 discuss a procedure for the design of a transport aircraft fueselage structure. The stress constraints are satisfied by using the FSD algorithm. The displacement constraint at the critical point is then satisfied by using the algorithm based on the mathematical optimality criterion discussed in Reference 27. A nonlinear mathematical programming algorithm is used for the detailed design of the panels to satisfy local buckling, system buckling, and allowable stresses.

Taig and Kerr in Reference 41 use a recurrence relation based on the mathematical optimality criterion for the displacement constraint problem in order to study the increased aeroelastic effectiveness of a taileron which is idealized with 1,358 elements and 540 nodes. The aeroelastic effectiveness target were converted into generalized displacement requirements. The Lagrange multipliers for the multiple constraints are evaluated by using a Newton-Raphson iterative methods.

Gellatly, Dupree, and Berke in Reference 20 discuss the development of the computer program OPTIM II for designing a minimum weight structure which satisfies stress and displacement constraints. This program is the extended version of the original OPTIM I program (Reference 27) which contained only bar elements. In OPTIM II the stress constraint problem, is solved by using the FSD concept, and the displacement constraint problem, is solved by using the recurrence relation based on the mathematical optimality criterion. For the multiple displacement constraint problem an envelope method is used to select the member size, where the largest area is selected for each element among all the loading and displacement combinations. In the case of a displacement constraint problem the stresses in the elements are satisfied by treating them as passive constraints. A single stress based on the von Mises criteria or a buckling stress based on a simple panel buckling equation is used for the plate elements. The use of the program is illustrated by the design of an idealized swept wing.

Dwyer in Reference 52 discusses the capabilities of the program ASOP-2 which is the improved and modified version of ASOP-1 (Reference 47). In ASOP-2 the displacement constraint algorithm established on the concept of designing a structure on the basis of a single critical constraint developed in References 12 and 13 is incorporated. The program also has a limited capability for sizing fiber reinforced orthotropic plate elements.

Khot, Venkayya, and Berke in Reference 53 present a method based on the optimality criterion for designing a fiber reinforced structure with stress and multiple displacement constraints. A modified recurrence relation based on the concept of a uniform strain energy density, that takes into consideration the differing stress limits in the elements, is used to design and idealize stress constrained wing structure with 100 nodes and 296 elements. The displacement constraint problem is solved using a recurrence relation based on the mathematical optimality criterion for the multiple constraint problem. The Lagrange multipliers are determined by using a recurrence relation derived from the constraint equations. For the displacement constraint problem the stress constraints are treated as passive constraints. A box-beam and a wing structure are designed with a multiple displacement constraint requirement.

Mathematical programming methods have usually been applied to more simplified models because of their computational cost. Haftka and Starnes (54-56), Schmit and Miura (21, 57), and Rizzi (58) present some typical examples of the application of mathematical programming methods to designing wing structures subject to stress and displacement constraints. The optimization procedures make use of approximation techniques to reduce the computational cost. The most complex example (Haftka and Starnes, Reference 56) is that of a high aspect ratio wing structure modeled by 290 finite elements with 146 independent design variables.

Schmit and Miura in References 21 and 29, discuss the development of a mathematical programming based computer program, ACCESS 1. This is an incore program which uses the optimization algorithms CONMIN (Reference 59) or NEWSUMT. CONMIN is a general purpose program that uses the method of feasible direction. The NEWSUMT algorithm employs an unconstrained minimization technique based on the extended interior penalty function. The problem is expressed in terms of the reciprocal design variable, and the concept of variable linking, a constraint deletion technique and an approximate analysis method are used to improve the efficiency of the algorithm. The design of a wing carry through structure obtained by using ACCESS 1 is compared with the results given in Reference 16 where the optimality criterion approach is used. A swept wing and a delta wing to satisfy the stress and displacement constraints are designed by using ACCESS 1.

Khot, Venkayya, and Berke in References 60 and 61, present a method based on using a recurrence relation derived from the mathematical optimality criterion for the design of a three-dimensional dome structure to satisfy system stability and stress constraints. The stability requirement is defined as a stiffness constraint and is measured by the Rayleigh quotient. The stress constraints are treated as passive constraints.

Rizzi in Reference 58 uses the recurrence relation based on the optimality criterion approach proposed in Reference 28 to design a wing box beam and a swept wing with stress and displacement constraints. The results are compared with those given in Reference 29. The Lagrange multipliers for the multiple constraint problems are evaluated by solving a set of linear equations involving the constraint gradients. The equations associated with the negative Lagrange multipliers of the inactive constraints are deleted during the solution. The equations are solved by using the Gauss-Seidel iterative scheme.

Isakson, Pardo, and Lerneer in Reference 22 (also see Reference 23) discuss the capabilities of the computer program ASOP-3, which is the modified and improved version of ASOP-2 (Reference 52). This program has the capability to design a structure idealized with up to 3,000 finite elements and 1,000 nodes. The stress constraint problem is solved by the FSD algorithm. The resizing of the elements for the deflection constraint (a single constraint or a linear combination of nodal deflections) is achieved by using a recurrence relation based on the mathematical optimality criterion which is different from the one used in ASOP-2. The capabilities of ASOP-3 are demonstrated by designing a cantilevered wing model and a bomber fin model.

Khot in Reference 62 discusses the capability of an incore optimization program OPTCOMP. This program can be used to optimize or analyze a metal or composite structure, which can be idealized by bar elements, shear panels and triangular or quadrilateral membrane plate elements. The structure can be optimized by using a recurrence relation based on the optimality criterion of uniform strain energy density or the FSD concept for different strength criteria selected by the user. The four strength criteria normally used for metal and composite structures are included. Membrane plates can be designed to prevent local buckling, and the element can be linked to have the same size if desired. In the analysis mode the program may be instructed to isolate these elements which do not satisfy the specified strength criteria. The use of the program is illustrated by the design of a box beam.

Khot, Venkayya, Berke, and Schrader in Reference 63 have developed a method to design a minimum weight fiber reinforced composite wing strucutre with stress and specific twist constraints. The method is based on a mathematical optimality criterion. A wing strucutre is designed to have a specific negative or positive twist corresponding to "wash out" and "wash in" conditions and also to satisfy the strength criteria for all the elements. The strength criteria in the elements are satisfied by considering the stress constraints as passive constraints.

Haftka and Starnes in Reference 56 present a quadratic extended interior penalty function mathematical programming method for the design of a wing structure subjected to stress, displacement, and minimum size constraints. The method is illustrated by designing statically loaded high- and low-aspect ratio wings.

Austin in Reference 64 proposes a method based on a mathematical optimality criterion to design a structure with multiple equality constraints. The method is illustrated by designing a composite intermediate complexity wing with multiple twist constraints. The starting design for the twist constraint problem is the strength based design obtained by the FSD procedure.

Schmit and Miura in Reference 57 review the capabilities of the program ACCESS-2, which is a modified version of ACCESS-1 (Reference 29). ACCESS-2 uses a new optimization module NEWSUMT2 and has the capability to consider thermal effects and to design fiber reinforced composite structures. The use of the program is illustrated by designing a delta wing (Reference 29) with a fiber reinforced composite skin.

Venkayya in Reference 65 compares the effect of using a recurrence relation based on the optimality criterion of uniform strain energy density with a FSD procedure on the material distribution of a fiber reinforced composite wing structure with stress constraints. The FSD design is obtained by using ASOP-3 (Reference 22), and the other design is obtained by using OPTSTAT (Reference 66).

Starnes and Haftka in Reference 67 extend the previous work of Reference 56 to include wing twist, panel buckling and fiber reinforced composite materials.

Sander and Fleury in Reference 43 present a mixed method using mathematical programming and the optimality criterion concept. An aircraft spoiler is designed to satisfy stress and displacement constraints. The displacement constraints on the trailing edge are such that the trailing edge has to remain straight within a specified tolerance under four loading conditions. The structure is first designed using the FSD algorithm, and the projection method of nonlinear programming is used to satisfy the displacement constraints.

Schmit and Ramanathan in Reference 68 discuss the modified version of ACCESS-1 (Reference 29) in which constraints on local buckling and system buckling are included. A delta wing structure with stress and local buckling constraints is designed to illustrate the method.

Adelman, Sawyer, and Shore in Reference 69 present a method based on the mixed approach which incorporates the optimality criterion and the mathematical programming method to optimize structures at an elevated temperature. The application of the method is illustrated by designing a wing box.

Schmit and Fleury in Reference 70 discuss the extension of the previous ACCESS-2 program (Reference 57) to include the optimization algorithm based on the dual method of nonlinear programming. In this algorithm the design variables are modified by using the recurrence relation based on the mathematical optimality criterion, and the Lagrange multipliers are calculated by maximizing the Lagrangian expressed as a function of the Lagrange multipliers by a mathematical programming procedure. A delta wing, previously studied in Reference 57, is designed by using this approach, and it is shown that this algorithm is more efficient than the ones used in the previous ACCESS series.

4.2.2 Buckling Constraints

While buckling constraints are sometimes encountered at the global level (see, for example, Simitses and Ungbhakorn, Reference 71) they are most commonly encountered at the local level for aerospace structures. That is, typically we are concerned with the buckling of individual panels on a wing or a fuselage structure rather than the global buckling of the entire structure. Because of the small number of design variables required for designing an individual panel, the panel design is ideally suited for mathematical programming techniques. Agarwal and Davis (72), Stroud and Agranoff (73 and 74), Schmit and Farshi (75), and Anderson and Stroud (76) present typical applications of general mathematical programming techniques to the design of panels subject to stress and buckling constraints.

Difficulties are encountered when an entire wing or fuselage structure which contains many individual panels has to be designed. The design is subject to global constraints such as displacement constraints and to local constraints such as stress and buckling. One approach (Starnes and Haftka, Reference 67) is to use a mathematical programming approach disregarding the local-global nature of the problem. However, this approach is very costly. For example, in that work, a simple wing model with 73 nodes, 413 finite elements, and 33 design variables required 20-25 minutes of CDC Cyber 173 CPU time. Other approaches require iterating between the local panel design and the global structural design. Sobieszczanski and Loendorf (19) use FSD for the global design and mathematical programming for the local panel design for fuselage structures. An example of a fuselage structural model consisting of about 1,900 degrees of freedom and 1,200 finite elements were used. Ramanathan (77) and Schmit and

Ramanathan (68) use a multilevel all mathematical programming approach to the local-global design problem.

4.2.3 Natural Frequency Constraints

Frequency constraints are used in aerospace applications not only when there are vibration and resonance problems but also as a means of controlling aerodynamic flutter problems. Because frequency constraints are global in nature they do not require a large number of design variables, and are therefore easily handled by mathematical programming algorithms. Kataria and Murthy (78), Schmit and Miura (57), Miura and Schmit (79), Schmit and Fleury (70), and Rao, et al., (80), discuss several applications of mathematical programming algorithms to the design of wing structures subject to frequency constraints. When it is required to optimize the structure simultaneously for frequency constraints and local constraints, such as stress constraints, a common approach is to use FSD in conjunction with an optimality criterion method. Taig and Kerr (41) for example, present the design of a fin and a rudder using this latter approach.

4.2.4 Thermal Constraints

Constraints on the temperature of a structure subjected to thermal load are usually handled by the designer of a thermal protection system rather than the structural analyst. However, in the past few years, it is becoming clear that in some cases the design of the structure can have a significant effect on the temperature distribution. In the design of reentry vehicles, like the space shuttle, it has been shown that a simultaneous design of the structure and its thermal protection system may be more efficient than a sequential approach (Adelman, et al., 69). As a result there has been recent interest of structural analysts in design under thermal constraints (Adelman and Sawyer, 81; Roe, et al., 82; and Haftka and Shore, 83).

4.2.5 Aeroelastic Constraints

Constraints on the aeroelastic behavior of structures are very important for the design of aerospace structure. The problem of minimum weight design subject to such constraints, and especially flutter constraints, has attracted a lot of interest. Because such constraints are global in nature, they do not require a large number of design variables and can be handled reasonably well by standard mathematical programming algorithms. It is surprising, therefore, that rather than finding papers on applications of such standard algorithms to practical design problems we find many papers that describe algorithms specialized to aeroelastic constraints along with their applications to highly simplified examples. Several survey papers are also available including McIntosh, et al., (84), Ashley, et al., (85), and Stroud (86). Stroud presents an excellent review of the state-of-the-art and the discussion here draws upon and updates the source.

The first class of specialized algorithms for design under flutter constraints includes mathematical programming techniques which use nonstandard search procedures--Rudisill and Bhatia, (87); Simodynes, (88, 89); and O'Connell and coworkers, (90, 91). A second class includes specialized optimality criterion algorithms developed for the flutter constraint--Triplett (92), Siegel (93), Pines and Newman (94, 95), Segenreich and McIntosh (96), Haftka and coworkers (97, 98), Wilkinson and coworkers (99, 100), and Lansing, et al., (101). There are also applictions of standard mathematical programming algorithms-- Stroud, et al., (102), Fox and coworkers (103-105), Rao (106), Haftka (107, 108), McCullers and Lynch (109, 110), Miura (111), Taylor and Gwin (112), Gwin and coworkers (113-116), Andries, et al., (117), Erbug (118), Phoa (119), and Chipman and Malone (120).

One problem that confronts the designer is that the flutter speed may be a discontinuous function of structural parameters. Some of the design procedures (109, 110, 121) overcome the problem by reformulating the flutter constraint to apply to the damping of the structural motion which is a continuous function of structural parameters.

Other aeroelastic constraints that have been considered in wing and control surface design are divergence speed (122, 123), control of effectiveness (124), and improved performance or "aeroelastic tailoring" (125-127).

4.2.6 Reducing the Number of Analyses

The high computational cost of structural optimization is mostly due to the large number of structural analyses required during the design process. Attempts to reduce this high cost are directed either at developing fast reanalysis techniques or to using derivatives of the constraints to approximate them during portions of the design process and thus obviate the need for reanalysis (e.g., 128, 129). Both of these approaches were pursued in applications to aeroelastic constraints. Fast reanalysis algorithms were developed by Bhatia (87) and Haftka and Yates (130, 131). Expressions for derivatives of flutter parameters or the eigen-modes which are needed for flutter calculations were derived by Rudisill and Bhatia (132, 133) and Rao (134). Design procedures incorporating approximation were developed by Austin (71) and Haftka and Prasad (108).

4.3 CONCLUSIONS

From this discussion, one sees that a number of different approaches based on either the nonlinear programming, the optimality criterion, or some combination of these have been developed by different investigators. Many of the early algorithms based on the nonlinear programming methods are not as efficient as some of the most recent ones. Some of the reasons for the inefficiency are: (1) the number of analyses of the structure is large, (2) all the constraints imposed on the structure are included in the mathematical model, and (3) the methods used to modify the design vector at each iteration are not efficient. The success and efficiency of the methods based on the mathematical optimality criterion have been attributed primarily to two facts: (1) only constraints which are expected to be critical in developing the algorithm are usually included; and (2) the recurrence relation used to modify the design variable is directly derived from the optimality criterion.

Finally, it should be noted that most of the papers in the literature make a distinction between algorithm based on the optimality criterion and the nonlinear programming methods. It is also generally indicated that algorithms based on the nonlinear mathematical programming methods are not as efficient as those based on the optimality criterion. However, it must be pointed out here that algorithms derived from the mathematical optimality criterion may also be developed by using the projection method (see Reference 33) or the dual method (see Reference 135) of nonlinear programming, and there is a relationship between these basic methods. Therefore, the important issue is not so much one of efficiency of mathematical programming, but one of choosing the best algorithm.

4.4 REFERENCES

1. Heyman, J., "Plastic Design of Beams and Frames for Minimum Material Consumption," Quart. Appl. Math., Vol. 8, 1951, pp. 373-381.

2. Livesley, R. K., "Optimum Design of Structural Frames for Alternative Systems of Loading," Civil Engr. and Public Works Review, Vol. 54, No. 636, June 1959, pp. 737-740.

3. Pearson, C. W., "Structural Design by High Speed Computing Machines," Proc. 1st Conference on Electronic Computation ASCE, New York, 1958, pp. 417-436.

4. Schmit, L. A., "Structural Design by Systematic Synthesis," Proc. 2nd Conference on Electronic Computation ASCE, New York, 1960, 105-122.

5. Schmit, L. A., Kicher, T. P., and Morrow, W. M., "Structural Synthesis Capability for Integrally Stiffened Waffle Plates," AIAA J., Vol. 1, No. 12, 1963, pp. 2820-2836.

6. Kicher, T. P., "Structural Synthesis of Integrally Stiffened Cylinders," J. of Spacecraft and Rockets, Vol. 5, No. 1, 1968, pp. 62-67.

7. Morrow, W. M. and Schmit, L. A., "Structural Synthesis of a Stiffened Cylinder," NASA CR-1217, 1968.

8. Stroud, W. J. and Skyes, N. P., "Minimum Weight Stiffened Shells with Slight Meridianal Curvature Designed to Support Axial Compressive Loads," AIAA J., Vol. 7, No. 8, 1969, pp. 1599 - 1601.

9. Chao, T., "Minimum Weight Design of Stiffened Fiber Composite Cylinders," AFML-TR-69-251, 1969.

10. Kicher, T. P. and Chao, T. L., "Minimum Weight Design of Stiffened Fiber Composition Cylinder," J. of Aircraft, Vol. 8, No. 7, 1971, pp. 562-568.

11. Stroud, W. J., Dexter, C. B., and Stein, M., "Automated Preliminary Design of Simplified Wing Structures to Satisfy Strength and Flutter Requirements," NASA TN D-6534, 1971.

12. Schmit, L. A. and Thornton, W. A., "Synthesis of an Airfoil at Supersonic Mach Number," NASA CR-144, 1965.

13. Miura, H., "An Optimal Configuration Design of Lifting Surface Type Structure under Dynamic Constraints," Ph.D. Dissertation, CWRU, Cleveland, Ohio, 1971.

14. Rao, S. S., "Automated Optimum Design of Aircraft Wings to Satisfy Strength, Stability, Frequency, and Flutter Requirements," Ph.D. Dissertation, CWRU, Cleveland, Ohio, 1971.

15. Fox, R. L., Miura, H., and Rao, S. S., "Automated Design Optimization of Supersonic Airplane Wing Structures Under Dynamic Constraints," AIAA Paper No. 72-333, April 1972.

16. Lynch, R. W. and McCullers, L. A., "Wing Aero-Elastic Synthesis Procedure," AFFDL-TR-73-111, March 1973.

17. Gwin, L. B. and Taylor, R. F., "A General Method for Flutter Optimization," AIAA J, Vol. 11, No. 13, 1973, pp. 1613-1617.

18. Giles, G. L., Blackburn, C. L., and Dixon, S. C., "Automated Procedures for Sizing Aerospace Vehicle Structures (SAVES)," Journal of Aircraft, Vol. 9, No. 12, 1972, pp. 812-819.

19. Sobieszczanski, J. and Loendorf, D., "A Mixed Optimization Method for Automated Design of Fuselage Structures," Paper presented at the 13th AIAA/ASME/SAE Structures, Structural Dynamics, and Materials Conference, San Antonio, Texas, 1972.

20. Gellatly, R. A., Dupree, D. M., and Berke, L., "OPTIM II: A Magic Compatible Large Scale Automated Minimum Weight Design Program," AFFDL-TR-97, Vol. I and II, July 1974.

21. Schmit, L. A. and Miura, H., "A New Structural Analysis/Synthesis Capability - ACCESS 1," AIAA Journal, Vol. 14, No. 5, pp. 661-671, May 1976.

22. Isakson, G. and Pardo, H., "ASOP-3: A Program for the Minimum-Weight Design of Structures Subjected to Strength and Deflection Constraints," AFFDL-TR-76-157, December 1976.

23. Isakson, G., Pardo, H., Lerner, E., and Venkayya, V. B., "ASOP-3: A Program for the Optimum Design of Metallic and Composite Structures Subjected to Strength and Deflection Constraints," AIAA Paper 77-378, presented at AIAA/ASME 18th Structures and Structural Dynamics and Materials Conference, San Diego, March 1977.

24. Gellatly, R. A., Berke, L., and Gibson, W., "The Use of Optimality Criteria in Automated Structural Design," Paper presented at the 3rd Conference on Matrix Methods in Structural Mech., WPAFB, Ohio, October 1971.

25. Venkayya, V. B., Khot, N. S., and Reddy, N. S., "Energy Distribution in an Optimum Structural Design," AFFDL-TR-68-156, March 1969.

26. Berke, L., "An Efficient Approach to the Minimum Weight Design of Deflection Limited Structures," AFFDL-TM-70-4-FDTR, May 1960.

27. Gellatly, R. A. and Berke, L., "Optimal Structural Design," AFFDL-TR-70-165, April 1970, Air Force Flight Dynamics Laboratory, Wright-Patterson Air Force Base, Ohio.

28. Kiusalaas, J., "Minimum Weight Design of Structures via Optimality Criteria," NASA TN D-7115, 1972.

29. Schmit, L. A. and Miura, H., "Approximation Concepts for Efficient Structural Synthesis," NASA CR-2552, March 1976.

30. Fleury, C. and Geradin, M., "Optimality Criteria and Mathematical Programming in Structural Weight Optimization," *J. Computers and Struct.* 8, 7-17, 1977.

31. Fleury, C. and Sander, G., "Relationships between Optimality Criteria and Mathematical Programming in Structural Optimization," *Proc. Symp. Applications of Computer Mech Engineering*, (ed. C. Wellford, Jr.), Univ. of Southern California, 507-520, 1977.

32. Khot, N. S., Berke, L., and Venkayya, V. B., "Minimum Weight Design of Structures by the Optimality Criterion and Project Method," AIAA 79-0720, Paper presented at AIAA/ASME/ASCE/AMS, 20th Structures, Structural Dynamics and Materials Conference, St. Louis, MO, April 1979.

33. Fleury, C. and Schmit, L. A., "Primal and Dual Methods in Structural Optimization," *Journal of the Structural Division*, ASCE, Vol. 106, No. ST5, Proc. Paper 15431, May 1980, pp. 1117-1133.

34. Lansing, W., Dwyer, W., Emerton, R., and Ranalli, E., "Application of Fully Stressed Design Procedures to Wing Empennage Structures," AIAA/ASME 11th Struct., Structural Dynamics and Materials Conf., Denver, Colorado (1970).

35. Lansing, W., Dwyer, W., Emerton, R., and Ranalli, E., "Application of Fully Stressed Design Procedures to Wing and Empennage Structures," *J. of Aircraft*, Vol. 8, No. 9, September 1971.

36. Giles, G. L. and McCullers, L. A., "Simultaneous Calculation of Aircraft Design Loads and Structural Member Sizes," AIAA Paper No. 75-965, 1975.

37. Adelman, H. M., Walsh, J. L., and Narayanaswami, R., "An Improved Method for Optimum Design of Mechanically and Thermally Loaded Structures," NASA TN D-7965, 1975.

38. Adelman, H. M. and Narayanaswami, R., "Resizing Procedure for Structures Under Combined Mechanical and Thermal Loading," *AIAA Journal*, Vol. 14, No. 10, 1976, pp. 1484-1486.

39. Gellatly, R. A., Gallagher, R. H., and Luberacki, W. A., "Development of a Procedure for Automated Synthesis of Minimum Weight Structures," AFFDL-TDR-64-141, October 1964.

40. Gellatly, R. A., "Development of Procedure for Large Scale Automated Minimum Weight Structural Design," AFFDL-TR-66-180.

41. Taig, I. C. and Kerr, R. I., "Optimization of Aircraft Structures with Multiple Stiffness Requirements," Second Symposium on Structural Optimization, AGARD-CP-123, Milan, Italy, April 1973.

42. Venkayya, V. B., Khot, N. S., and Berke, L., "Application of Optimality Criteria Approaches to Automated Design of Large Practical Structures," Second Symposium on Structural Optimization, AGARD-CP-123, Milan, Italy, April 1973.

43. Sander, G. and Fleury, C., "Mixed Method in Structural Optimization," Int. J. for Num. Met. In Engineering, Vol. 13, No. 2, December 1978.

44. Melosh, R. J. and Luik, R., "Approximate Multiple Configuration Analysis and Allocation for Least Weight Structural Design," AFFDL-TR-67-59, April 1967.

45. Johnson, J. R., Melosh, R. J., and Luik, R., "Optimum Structural Design," 25th Meeting of the Structures and Materials Panel of AGARD, September 25-29, 1967.

46. Dwyer, W., Rosenbaum, J., Shulman, M., and Pardo, H., "Fully Stressed Design of Air Frame Redundant Structures," Proceedings of the Second Conference on Matrix Methods in Structural Mechanics, AFFDL-TR-68-150, December 1969.

47. Dwyer, W. J., Emerton, R. K., and Ojalvo, I. U., "An Automated Procedure for the Optimization of Practical Aerospace Structures. Vol. I - Theoretical Development and User's Information, Vol. II -Programmer's Manual," AFFDL-TR-70-118, April 1971.

48. Lansing, W., Dwyer, W., Emerton, R., and Ranalli, E., "Application of Fully Stressed Design Procedures to Wing and Empennage Structures," J. Aircraft, Vol. 8, pp. 683-688, September 1971.

49. Khot, N. S., Venkayya, V. B., Johnson, C. D., and Tischler, V. A., "Optimum Design of Advanced Composite Structures," Paper presented at the Sixth St. Louis Symposium on Composite Materials in Engineering Design, St. Louis, May 11-12, 1972.

50. Khot, N. S., Venkayya, V. B., Johnson, C. D., and Tischler, V. A., "Optimization of Fiber Reinforced Composite Structures," Int. J. Solids Structures, Vol. 9, pp. 1225-1236, Pergamon Press, September 1973.

51. Sobieszczanski, J., "Sizing of Complex Structures by the Integration of Several Different Optimal Design Algorithms," AGARD Lecture Series No. 70, Structural Optimization, 1974.

52. Dwyer, W. J., "An Improved Automated Structural Optimization Program," AFFDL-TR-75-96, September 1974.

53. Khot, N. S., Venkayya, V. B., and Berke, L., "Optimum Design of Composite Structures with Stress and Deflection Constraints,"

AIAA Paper No. 75-141, presented at AIAA 13th Aerospace Sciences Meeting, Pasadena, California, January 20-22, 1975.

54. Haftka, R. T. and Starnes, J. H., Jr., "WIDOWAC (Wing Design Optimization with Aeroelastic Constraints): Program Manual," NASA TM X-3071, 1974.

55. Haftka, R. T. and Starnes, J. H., Jr., "Application of a Quadratic Extended Interior Penalty Function for Structural Optimization," AIAA Paper 75-764, AIAA/ASME/SAE 16th Structures, Structural Dynamics, and Material Conference, Denver, Colorado, May 1975.

56. Haftka, R. T. and Starnes, J. H., Jr., "Applications of a Quadratic Extended Interior Penalty Function for Structural Optimization," AIAA Journal, Vol. 14, No. 6, 1976, pp. 718-724.

57. Schmit, L. A. and Miura, H., "An Advanced Structural Analysis/Synthesis Capability - ACCESS 2," International Journal for Numerical Methods in Engineering, Vol. 12, 1978, pp. 353-357.

58. Rizzi, P., "Optimization of Multi-Constrained Structures Based on Optimality Criteria," AIAA/ASME/SAE 17th Structures, Structural Dynamics and Materials Conference, King of Prussia, PA, May 1976.

59. Vanderplaats, G. N., "CONMIN - A Fortran Program for Constrained Function Minimization: User's Manual," NASA TM X-62,282, 1973.

60. Khot, N. S., Venkayya, V. B., and Berke, L., "Optimization of Structures for Strength and Stability Requirements," AFFDL-TR-73-98, December 1973.

61. Khot, N. S., Venkayya, V. B., and Berke, L., "Optimum Structural Design with Stability Constraints," International Journal for Numerical Methods in Engineering, Vol. 10, October 1976, pp. 1097-1114.

62. Khot, N. S., "Computer Program (OPTCOMP) for optimization of Composite Structures for Minimum Weight Design," AFFDL-TR-76-149, February 1977.

63. Khot, N. S., Venkayya, V. B., Berke, L., and Schrader, K., "Optimum Design of Composite Wing Structures with Twist Constraint for Aeroelastic Tailoring," AFFDL-TR-76-117, December 1976.

64. Austin, F., "A Rapid Optimization Procedure for Structures Subjected to Multiple Constraints," Paper No. 77-374, AIAA/ASME/SAE 18th Structures, Structural Dynamics and Materials Conference, San Diego, California, March 1977.

65. Venkayya, V. B., "Structural Optimization: A Review and Some Recommendations," International Journal of Numerical Methods in Engineering, Vol. 13, No. 2, December 1978.

66. Venkayya, V. B. and Tischler, V. A., "OPTSTAT - A Computer Program for Optimal Design of Structures Subjected to Static Loads," AFFDL-TR-80- (in preparation).

67. Starnes, J. H. and Haftka, R. T., "Preliminary Design of Composite Wing for Buckling, Strength and Displacement Constraints," *Journal of Aircraft*, Vol. 16, pp. 564-570, 1979.

68. Schmit, L. A., Jr. and Ramanatham, R. K., "Multilevel Approach to Minimum Weight Design Including Buckling Constraints," *AIAA Journal*, Vol. 16, No. 1, 1978, pp. 97-104.

69. Adelman, H. M., Sawyer, P. L., and Shore, C. P., "Development of Methodology for Optimum Design of Structures at Elevated Temperatures," AIAA Paper 78-469, presented at 19th Structures, Structural Dynamics and Materials Conference, Bethesda, MD, April 1978.

70. Schmit, L. A., Jr. and Fleury, C., "An Improved Analysis/Synthesis Capability Based on Dual Methods - ACCESS 3," AIAA Paper 79-0721, presented at 20th Structures, Structural Dynamics and Materials Conference, St. Louis, MO, April 1979.

71. Simitses, George J. and Ungbhakorn, Variddhi, "Weight Optimization of Stiffened Cylinders Under Axial Compression," *Computers and Structures*, Vol. 5, pp. 305-314, 1975.

72. Agarwal, B. and Davis, R., "Minimum-Weight Designs for Hot-Stiffened Composite Panels Under Uniaxial Compression," NASA TN D-7779, 1974.

73. Stroud, W. J. and Agranoff, N., "Minimum-Mass Design of Filamentary Composite Panels Under Combined Loads: Design Procedure Based on Simplified Buckling Equations," NASA TN D-8257, 1976.

74. Stroud, W. J., Agranoff, N., and Anderson, M. S., "Minimum-Mass Design of Filamentary Composite Panels Under Combined Loads: Design Procedure Based on a Rigorous Buckling Analysis," NASA TN D-8417, 1977.

75. Schmit, L. A., Jr. and Farshi, B., "Optimum Design of Laminated Fiber Composite Plates," *International Journal for Numerical Methods in Engineering*, Vol. 11, 1977, pp. 623-640.

76. Anderson, M. S. and Stroud, W. M., "A General Panel Sizing Computer Code and Its Application to Composite Structural Panels," AIAA Paper 78-467, 1978.

77. Ramanathan, R. K., "A Multilevel Approach in Optimum Design of Structures Including Buckling Constraints," Ph.D. Thesis, University of California, Los Angeles, 1976.

78. Katarya, R. and Murthy, P. M., "Optimization of Multicell Wings for Strength and Natural Frequency Requirements," *Computers and Structures*, Vol. 5, No. 4, 1975, pp. 225-232.

79. Miura, H. and Schmit, L. A., Jr., "Second Order Approximation of Natural Frequency Constraints in Structural Synthesis," *International Journal of Numerical Methods in Engineering*, Vol. 13, No. 2, 1978, pp. 203-228.

80. Rao, V. R., Iyengar, N. G. R., and Rao, S. S., "Optimization of Wing Structures to Satisfy Strength and Frequency Constraints," *Computers and Structures*, Vol. 10, pp. 669-677, 1979.

81. Adelman, H. M. and Sawyer, P. L., "Inclusion of Explicit Thermal Requirements in the Optimum Design of Structures," NASA TM X-74017 (1977).

82. Rao, G. V., Shore, C. P., and Narayanaswami, R., "An Optimality Criterion for Resizing Heated Structures with Temperature Constraints," NASA TN D-8285 (1977).

83. Haftka, R. H. and Shore, C. P., "Approximation Methods for Combined Thermal/Structural Design," NASA TP-1428, 1979.

84. McIntosh, S. C., Weisshaar, T. A., and Ashley, H., "Progress in Aeroelastic Optimization - Analytical Versus Numerical Approaches," AIAA Struct., Dyn., and Specialists Conference, New Orleans (1969).

85. Ashley, H., McIntosh, S. C., Jr., and Weatherill, W. H., "Optimization Under Aeroelastic Constraints," Chap. 11 of *Structural Design Applications of Mathematical Programming Techniques*. AGARDograph No. 149, Edited by G. G. Pope and L. A. Schmit, 1971.

86. Stroud, W. Jefferson, "Automated Structural Design With Aeroelastic Constraints: A Review and Assessment of the State-of-the-Art," ASME Symposium on Structural Optimization, 1974, Same Winter Annual Meeting, New York, NY, November 17-21, 1974.

87. Bhatia, K. G., "Rapid Iterative Reanalysis for Automated Design," NASA TN D-7357, October 1973.

88. Simodynes, E. E., "Gradient Optimization of Structural Weight for Specified Flutter Speed," AIAA Paper No. 73-390, AIAA/ASME/SAE 14th Structures, Structural Dynamics, and Materials Conference, Williamsburg, VA, March 1973.

89. Simodynes, E. E., "Gradient Optimization of Structural Weight for Specified Flutter Speed," *Journal of Aircraft*, Vol. 11, No. 3, 1974, pp. 143-147.

90. O'Connell, R. P., "Incremented Flutter Analysis," *J. Aircraft*, Vol. 11, No. 4, April 1974.

91. O'Connell, R. F., Radovcich, N. A., and Hassig, H. J., "Structural Optimization with Flutter Speed Constraints Using Maximized Step Size," *Journal of Aircraft*, Vol. 14, No. 1, 1977, pp. 85-89.

92. Triplett, W. E. and Ising, K. D., "Computer Aided Stabilator Design Including Aeroelastic Constraints," *Journal of Aircraft*, Vol. 8, No. 7, July 1971, pp. 554-561.

93. Siegel, S., "A Flutter Optimization Program for Aircraft Structural Design," AIAA Paper No. 72-795, AIAA Fourth Aircraft Design, Flight Test, and Operations Meeting, Los Angeles, CA, August 1972.

94. Pines, S. and Newman, M., "Structural Optimization for Aeroelastic Requirements," AIAA Paper No. 73-389, AIAA/ASME/SAE 14th Structures, Structural Dynamics, and Materials Conference, Williamsburg, VA, March 1973.

95. Pines, S. and Newman, M., "Constrained Structural Optimization for Aeroelastic Requirements," *J. Aircraft*, Vol. 11, No. 6, June 1974.

96. Segenreich, S. A. and McIntosh, S. C., "Weight Minimization of Structures for Fixed Flutter Speed via an Optimality Criterion," Proc. AIAA/ASME/SAE 16th Struct., Structural Dynmaics, and Materials Conference, Denver, Colorado (1975).

97. Haftka, R. T., Starnes, J. H., Jr., and Barton, F. W., "A Comparison of Two Types of Structural Optimization Procedures to Satisfy Flutter Requirements," AIAA Paper No. 74-405, AIAA/ASME/SAE 15th Structures, Structural Dynamics, and Materials Conference, Las Vegas, Nevada, April 1974.

98. Haftka, R. T., Starnes, J. H., Jr., Barton, F. W., and Dixon, S. C., "Comparison of Two Types of Structural Optimization Procedures for Flutter Requirements," *AIAA Journal*, Vol. 13, 1975, pp. 1333-1339.

99. Wilkinson, K., Lerner, E., and Taylor, R. F., "Practical Design of Minimum-Weight Aircraft Structures for Strength and Flutter Requirements," *J. Aircraft*, 13, 614-624 (1976).

100. Wilkinson, K., Markowitz, J., Lerner, D. George, and Batill, S. M., "FASTOP: A Flutter and Strength Optimization Program for Lifting-Surface Structures," *J. Aircraft*, 14, 581-587 (1977).

101. Lansing, W., Lerner, E., and Taylor, R. F., "Applications of Structural Optimization for Strength and Aeroelastic Design Requirements," Paper presented at the 45th AGARD Struct. and Materials Panel Meeting, Voss, Norway, AGARD Report No. 664 (1978).

102. Stroud, W. J., Dexter, C. B., and Stein, M., "Automated Preliminary Design of Simplified Wing Structures to Satisfy

Strength and Flutter Requirements," NASA TN D-6534, December 1971.

103. Fox, R. L., Miura, H., and Rao, S. S., "Automated Design Optimization of Supersonic Airplane Wing Structures Under Dynamic Constraints," AIAA Paper No. 72-333, AIAA/ASME/SAE 13th Structural Dynamics and Materials Conference, San Antonio, TX, April 1972.

104. Fox, R. L., Miura, H., and Rao, S. S., "Automated Design Optimization of Supersonic Airplane Wing Structures Under Dynamic Constraints," Journal of Aircraft, Vol. 10, No. 6, 1973, pp. 321-322.

105. Fox, R. L., Miura, H., and Rao, S. S., "Automated Design Optimization of Supersonic Airplane Wing Structures Under Dynamic Constraints," NASA CR-112319, 1973.

106. Rao, S. S., "Automated Optimum Design of Aircraft Wings to Satisfy Strength, Stability, Frequency and Flutter Requirements," Rep. No. 49, Div. Solid Mech., Struct., and Mech. Design, Case Western Reserve Univ., October 1971.

107. Haftka, R. T., "Automated Procedure for Design of Wing Structures to Satisfy Strength and Flutter Requirements," NASA TN D-7264, 1973.

108. Haftka, R. T. and Prasad, B., "Programs for Analysis and Resizing of Complex Structures," Computers and Structures, Vol. 10, 1979, pp. 323-330.

109. McCullers, L. A. and Lynch, R. W., "Composite Wing Design for Aeroelastic Requirements," Proceedings of the Conference of Fibrous Composites in Flight Vehicle Design, AFFDL-TR-72-130, U.S. Air Force, September 1972, pp. 951-972. (Available from DDC as AD 907 042L.)

110. McCullers, L. A. and Lynch, R. W., "Dynamic Characteristics of Advanced Filamentary Composite Structures: Vol. II, Aeroelastic Synthesis Procedure Development" AFFDL-TR-73-111, 1973.

111. Miura, H., "An Optimal Configuration Design of Lifting Surface Type Structures Under Dynamic Constraints," Rep. No. 48, Div. Solid Mech., Struct., and Mech. Design, Cast Western Reserve Univ., October 1971.

112. Taylor, R. F. and Gwin. L. B., "Application of a General Method for Flutter Optimization," Proceedings of the Second Symposium on Structural Optimization, AGARD-CP-123, April 1973, pp. 13-1 - 13-13.

113. Gwin, L. B. and McIntosh, S. C., Jr., "A Method of Minimum-Weight Synthesis for Flutter Requirements, Part I - Analytical Investigation," AFFDL-TR-72-22, Part I, June 1972.

114. Gwin, L. B. and McIntosh, S. C., Jr., "A Method of Minimum-Weight Synthesis for Flutter Requirements, Part II - Program Documentation," AFFDL-TR-72-22, June 1972.

115. Gwin, L. B. and Taylor, R. F., "A General Method for Flutter Optimization," AIAA Paper No. 73-391, AIAA/ASME/SAE 14th Structures, Structural Dynamics, and Materials Conference, Williamsburg, VA, March 1973.

116. Gwin, L. B. and Taylor, P. R., "A General Method for Flutter Optimization," AIAA Journal, Vol. 11, No. 12, December 1973, pp. 1613-1617.

117. Andries, R. A., Batill, S. M., and Taylor, R. R., "Demonstration and Application of a Minimum-Weight Synthesis Procedure for Flutter Requirements," AFFDL-TM-73-19-FYS, 1973.

118. Erbug, I. O., "Application of a Gradient Projection Technique to Minimum-Weight Design of Lifting Surfaces with Aeroelastic and Static Constraints," the Texas Institute for Computational Mechanics, TICOM Report 74-3 (1974).

119. Phoa, Y. T. and Chi, F. H., "Application of a Rand-Developed Nonlinear Programming Method to Flutter Optimization," Computers and Structures, Vol. 5, 1976, pp. 305-312.

120. Chipman, R. R. and Malone, J. B., "Application of Optimization Technology to Wing Store Flutter Prediction," AIAA Paper 77-453, 1977.

121. Haftka, R. T., "Parametric Constraints with Application to Optimization for Flutter Using a Continuous Flutter Constraint," AIAA Journal, Vol. 13, No. 4, 1975, pp. 471-475.

122. Gwin, L. B., "Optimal Aeroelastic Design of an Oblique Wing Structure," AIAA Paper No. 74-349, AIAA/ASME/SAE 15th Structures, Structural Dynamics, and Materials Conference, Las Vegas, NV, April 1974.

123. Gwin, L. B., "Optimal Aeroelastic Design of an Oblique Wing Structure," Journal of Aircraft, Vol. 13, No. 5, 1976, pp. 364-368.

124. Lerner, E. and Markowitz, "An Efficient Structural Resizing Procedure for Meeting Static Aeroelastic Design Objectives," AIAA/ASME 19th SDM Conf., Bethesda, MD, pp. 59-66 (1978).

125. Lynch, R. W. and Rogers, W. A., "Aeroelastic Tailoring of Composite Materials to Improve Performance," Proc. AIAA/ASME/SAE 17th Structures, Structural Dynamics and Materials Conference, Valley Forge, PA (1976).

126. Haftka, R. T., "Optimization of Flexible Wing Structures Subject to Strength and Induced Drag Constraints," AIAA Journal, Vol. 15, No. 8, 1977, pp. 1101-1106.

127. Gimmestad, D., "An Aeroelastic Optimization Procedure for Composite High Aspect Ratio Wings," AIAA Paper 79-726, 20th AIAA/ASHE/ASCE/AHS Structures, Structural dynamics, and Material Conference, St. Louis, MO, April 1979.

128. Schmit, L. A. and Farshi, B., "Some Approximation Concepts for Structural Synthesis," AIAA Paper No. 73-341, AIAA/ASME/SAE 14th Structures, Structural Dynamics, and Materials Conference, Williamsburg, VA, March 1973.

129. Schmit, L. A., Jr. and Farshi, B., "Some Approximation Concepts for Structural Synthesis," AIAA Journal, Vol. 12.

130. Haftka, R. T. and Yates, E. C., Jr., "On Repetitive Flutter Calculations in Structural Design," AIAA 12 Aerospace Sciences Meeting, Washington, DC, January 1974, Paper No. 74-141.

131. Haftka, R. T. and Yates, E. C., "Repetitive Flutter Calculations in Structural Design," Journal of Aircraft, Vol. 13, pp. 454-461, 1976.

132. Rudisill, C. S. and Bhatia, K. G., "Optimization of Complex Structures to Satisfy Flutter Requirements," AIAA Journal, Vol. 9, No. 8, 1971, pp. 1487-1491.

133. Rudisill, C. S. and Bhatia, K. G., "Second Derivatives of the Flutter Velocity and the Optimization of Aircraft Structures," AIAA Journal,, Vol. 10, No. 12, December 1972, pp. 1569-1572.

135. Rao, S. S., "Rates of Change of Flutter Mach Number and Flutter Frequency," AIAA Journal, Vol. 10, No. 11, November 1972, pp. 1526-1528.

136. Gellatly, R. A., Gallagher, R. H., and Luberacki, W. A., "Development of a Procedure for Automated Synthesis of Minimum Weight Structures," AFFDL-TDR-64-141, October 1964.

5. AN OUTLOOK FOR BASIC RESEARCH

5.1 INTRODUCTION

It is the purpose of this chapter to comment upon the state-of-the-art of research within structural optimization. Since structural optimization is at least formally in the realm of mathematical programming, it is natural to look to mathematical programming first for indications of future developments. In so doing, it is found that, from the aspect of algorithms, a rather dreary, dormant situation had existed, resulting in engineers attempting to develop their own algorithms in the absence of more official assistance. Out of this has come a strong school of structural optimization.

The state-of-the-art of mathematical programming has a close parallel in the state-of-the-art of linear systems of equations some 15 or so years ago. At that time, an engineer who wished to solve a large system of equations probably had to involve himself in the details of coding an algorithm or at least parts of an algorithm to do so. And incidentally, these efforts resulted in what is now called sparse matrix technology. In mathematical programming today, the demands far exceed the capabilities and the engineer is attempting to help out with algorithms as best he can. Predicting the outcome of these efforts at the present time would be premature.

There is another side to the participation of structural engineers in the creation of mathematical programming algorithms. This tends to involve the use of properties which are peculiar to structures and can have important ramifications. Returning again to the parallel with systems of linear equations, writing a general linear equation solver can be a major task while writing a solver for structures which are sparse and well-conditioned is a relatively simple task. Similarly, writing any **general** mathematical programming algorithms can be a major undertaking while writing an algorithm for a restricted class of structural optimization problems is likely to be much easier.

Structural optimization spans both general algorithmic discussions of mathematical programming and specific applications in structures which are restricted in scope. In the probable absence of major algorithmic advances in mathematical programming in the near future, structural engineers are likely to see structural optimization developing around properties peculiar to structures as a matter of default. This is the basic thread from which this chapter is constructed.

Another view of this situation comes from the existing literature. Figure 1, which has been taken from the list of books on mathematical programming and optimization (sections 8.1 and 8.2) indicates that there was no immediate response to the very remarkable effort of Arrow and a few others in 1958, but something of a steady buildup in the late 1960's followed by another peak in the mid-1970's which appears to respond to the earlier peak. By comparison (Figure 1), books on structural optimization (see Sections 8.1 and 8.2) are probably trailing the general optimization literature by approximately 5 years.

-55-

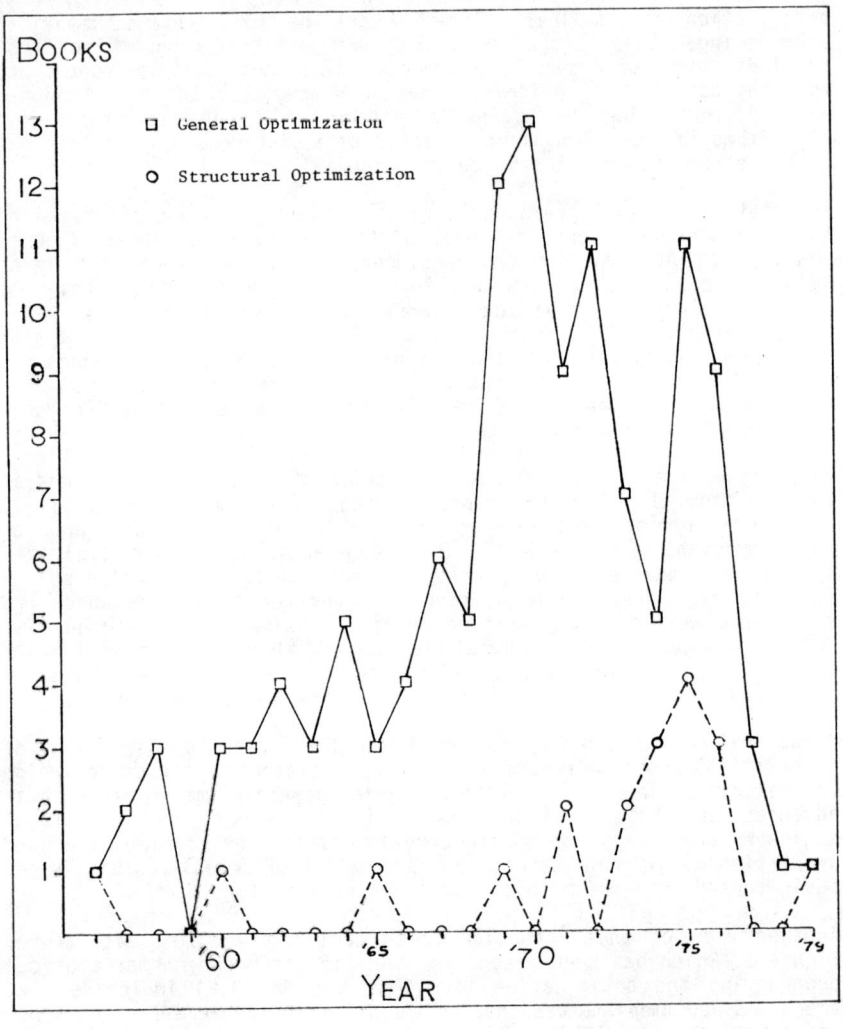

Figure 1. Books on Optimization Since 1956.

Another factor to be considered is, of course, hardware. Unbelievably, the cost of computing power continues to drop and the possibility (or probability) of ultracomputers is looming in the near future. It is quite conceivable that hardware developments will have a major impact on the practice of optimization. For example, efforts now being made to simplify structural optimization problems could be wasted in the long run by major development in hardware, which would make simple but extremely general procedures, such as piecewise linear programming, competitive again.

5.2 MATHEMATICAL PROGRAMMING ALGORITHMS

This discussion will be limited to the more general aspects of mathematical programming and not consider the more esoteric topics such as stochastic programming. Furthermore, in spite of the fact that much of structural optimization is concerned with discrete member sizes, little attention will be given here to integer programming due to its inherent combinatorial difficulties. In the following discussion, it will be assumed that either a continuous spectrum of member sizes is available or that the discrete spectrum can be adequately approximated by a continuous spectrum.

The reader is referred to Chapter 2 for formulation and methods of mathematical programming. As indicated earlier, the discipline of mathematical programming has stabilized and there are now many excellent reviews of the state-of-the-art available, such as References 1 through 6. It is the intention here to attempt only to characterize this actively.

It is the sense of References (1-3) and the sense of this report that there are no algorithmic surprises on the immediate horizon. For example, Reference 3 contains two sections on algorithmic advances which are concerned with refinements of existing algorithms rather than the development of new approaches. In fact, considerable interest today seems to lie with evaluating the performance of existing algorithms which clearly represents a stabilization of the state-of-the-art. As an example, an excellent paper by Ragsdell (7), compared data on existing nonlinear programming algorithms, to guide users in their selection.

References 1, 2, and 3 discuss these methods in some detail and there is little reason to repeat that discussion here. The hard fact of the matter is that demands far exceed the capabilities of existing nonlinear programming algorithms. With the exception of situations in which there are either a small number, (approximately 10), of independent variables or a particular simple (linear, quadratic, etc.) problem statement, the engineer cannot expect to find a "black box" to solve his optimization problem. This by no means implies that the engineer can not construct such a box himself; on the contrary.

5.3 STRUCTURAL OPTIMIZATION

There are, of course, both a theory and a practice of mathematical programming. Distinct from some areas of analysis, the practice of mathematical programming can allow a liberal dose of intuition since any process which results in a decreasing objective function is valid, no matter what its motivation.

Structural optimization, of course, spans both the theory and the practice of mathematical programming. Some of it, as mentioned above, follows directly as applications of mathematical programming theory. There is also an interesting side of structural optimization (of considerable practical importance) in which intuition plays a major role in developing algorithms. This section will attempt a brief look at the entire spectrum and attempt to show that (a) the intuition methods are simply the more formal methods with certain features omitted and (b) the hope for structural optimization in the near future lies with methods developed using the intuitive approach.

The structural optimization which follows directly as an application of mathematical progrmaming techniques has been discussed regularly in the literature and an excellent summary has recently been prepared by Jones (1). In gross terms again, the problem is simply one of finding an algorithm capable of dealing with a large number of variables since a large number of design variables is the basic fact of life of design. If an algorithm works, the designer is more likely to be interested in the details of his structure than in the details of the algorithm solving his problem for him. Unfortunately, with respect to mathematical programming, we are today in some nether land where demands exceed capabilities and engineers who wish to solve problems are in many cases forced to concern themselves with the algorithm to be used. This dilemma can of course have useful side effects.

There is an appealing optimization procedure for structures which involves a kind of analysis redesign sequence.

$$K^{(n)}\delta^{(n+1)} = P \quad \text{with} \quad K^{(n+1)} = F(\delta^{(n+1)}) \tag{5.1}$$

and

$$W^{(n+1)} \leq W^{(n)}$$

Roughly, a discrete system is represented in Equation (5.1) by the displacement formulation

$$K\delta = P \tag{5.2}$$

where

    K - stiffness matrix
    δ - joint displacement matrix
    P - joint load matrix

Given the member stiffnesses k_i, the stiffness matrix K is determined and so is the system response which is represented here by the displacement matrix δ. Equation (5.1) then implies that given the stiffness matrix at step n, the response at step n+1 (i.e., $\delta^{(n+1)}$) can be computed; given a **resizing rule** F new stiffness are computed such that the weight W decreases from step to step.

There are many ways to interpret Equation (5.1); certainly

$$K^{(n+1)} - K^{(n)} = F(\delta^{(n+1)}) - F(\delta^{(n)}) \qquad (5.3)$$

represents a feasible direction. Probably the ends of the spectra of interpretation lie in References (8-9). Venkayya, et al., (8) have developed a comprehensive set of procedures, they refer to as "optimality criteria" methods from a hybrid of considerations starting first of all with the partial Kuhn-Tucker conditions and finally using a good deal of intuition to arrive at a resizing rule F. They take several shortcuts in obtaining this rule and tend not to include all constraints specifically in their problem statement.

The other end of the spectrum is represented by Spillers (9) who considers much more simple problems but tends to remain within the theory of optimization.

In any case, the motivation is the same. When stuck in some computational morass, it is natural to abandon more general procedures and look to the peculiarities of the specific problem in hand. This has the double advantage of simplification and also fostering an increased understanding of structures.

The basic effectiveness of the approach outlined in Equation (5.2) lies in the fact that it is based in linear analysis. One of the most powerful technologies to develop in the last 15 years is our capability to deal with large sparse systems. An optimization technology developed on top of this analysis capability then has an enormous jump on any other methods which must start from scratch.

There is no parallel in optimization--even in linear programming--for this efficient technology now available to solve discrete linear systems. It is a happy circumstance that this technology carries over into structural optimization through the formulation of Equation (5.2) but this is clearly not the general case. The problem lies in the resizing rule which may become quite complex and in some cases dwarf the analysis segment of the algorithm.

5.4 HARDWARE

In Reference 10, Schwartz gives an interesting picture of some of the possibilities with respect to hardware in the near future. What he is saying is that it appears that we will shortly have on a single chip and at a low cost what we now call a large computer. The question is how to best make use of this forthcoming technology.

Schwartz predicts that we will soon have 10^7 to 10^8 transistors on a single chip which would give one chip more function than today's largest computers. The question is then how to combine these chips or modules into a high performance system and what should the software be. This is the basic question of V L S I (very large scale integration).

More than solving specific problems or making specific suggestions, Schwartz really raises a set of issues:

1. Current programming techniques are invalid for VLSI. Since current programming techniques and, in fact, current problem solving methods are serial in nature, they would waste the potential of VLSI.

2. What should be expected from VLSI and how should it be used? Schwartz suggests that (a) VLSI could be used to alleviate existing problems of software production and (b) new techniques of parallel or concurrent computations should be developed.

3. How should an infinite supply of small computers best be connected? There is a necessary trade-off to be made here between optimal communication between modules and problem of wiring.

4. What parallelism now exists in physical problems which can be exploited in VLSI? It is obvious that in order to exploit the potential of a large modular computer it will be necessary to deal with physical systems rather than individual physical problems. The structural design environment is a typical case in point.

5. What is an appropriate programming language for VLSI? Clearly it must be a very high level language beyond anything now available.

6. What are the theoretical limits? In order to develop the potential of VLSI properly we need to know something about what can be expected in an optimal situation. Computer theorists will have to come up with results of this kind.

The implications of VLSI in the areas of structural optimization or even mathematical programming are unknown at the moment. VLSI itself implies very cheap large scale computing. As mentioned earlier, cheap computing power may lead to a resurgence of simple but basic algorithms like sequential linear programming changing the state-of-the-art of structural optimization drastically. In any case, the new computer-on-a-chip world will probably not impact the general optimization computing environment for some years.

5.5 REFERENCES

1. Mathematical Programming Applications in Civil Engineering Design: Current Status and Future Trends (a collection of three articles compiled by Jones, H. L.), preprint 2897, ASCE, Spring Convention and Exhibit, Dallas, Texas, April 25-29, 1977.

2. Arora, J. S. and Haug, E. J., "Efficient Hybrid Methods of Optimal Structural Design," Proc. ASCE, EM3, June 1978, pp. 663-680.

3. White, W. W. (ed.), Computers and Mathematical Programming, Proc. Bicentennial Conf. on Math. Programming held at the National Bureau of Standards, Gaithersburg, Maryland, November 29 - December 1, 1976, NBS Special Publication 502, February 1978.

4. Himmelblau, D., Applied Nonlinear Programming, McGraw Hill, 1972.

5. Zoutendijk, G., Methods of Feasible Directions, Elseview Publ. Co., Amsterdam, 1960.

6. Fiacco, A. V. and McCormick, G. P., Nonlinear Programming: Unconstrained Minimization Techniques, Wiley, 1968.

7. Ragsdell, K. M., "On Some Experiments Which Delimit The Utility of Nonlinear Programming Algorithms," a paper presented at the ORSA/TIMS meeting in Los Angeles, California, November 13-15, 1978.

8. Venkayya, V. B., "Structural Optimization: A Review and Some Recommendations," International Journal for Numerical Methods in Engineering, 13, 1978, pp. 203-228.

9. Spillers, W. R., Interactive Structural Design, N. Holland, Amsterdam, 1975.

10. Schwartz, J. T., "Ultra-Computers," N. Y. University Report, 1978.

6. CONCLUSIONS

6.1 THE STATE-OF-THE-ART

The dominant topic in the recent literature on structural optimization concerns algorithms and techniques. Among these are aspects of selection, convergence, and economy.

The majority of problems investigated are theoretical and dimensionally small. Typically trusses and other simple structures are considered. The extension to larger practical design problems is slow and complex due to uncertainty in convergence and globality of optimum and the high cost of tackling such difficulties. For these reasons, fully stressed design appeals to many designers because it ignores most of these problems and the number of iterations may be limited at will. Sections 6.1.2 and 6.1.3 summarize recent experience concerning aspects of algorithm efficiency and fully stressed design.

Problems of shape and topology optimization are still considered very complex and the practical applications of the theoretical progress made quite rare. Substantial progress has been made in optimization of structures under dynamic loads.

The last few years have seen some consolidation in the sense that fruitful research has been made to compare and relate mathematical programming methods with optimality criterion methods. Furthermore, combinations of techniques were utilized to accelerate convergence and reduce cost.

Much of the optimization research has been concerned, thus far, with improved redesign algorithms to decrease computation time. Some of the effort may have been misdirected since electronic developments have drastically reduced the computer costs to the point where the main costs are often program development and use of peripheral devices. Thus speed and availability of large core size computers is no longer costly so increased computational efficiency may not be a concern except for very large structural systems.

A wide gap exists betwen numberical and analytical optimization methods. The interaction between the two groups of researchers of these methods is frequently lacking. While it is expected that computer applications would trail theoretical research, this lack of communication is rather disturbing, since progress is quite heavily dependent on the interaction between the two.

In spite of these difficulties optimization is being increasingly applied to design problems. The limited success of optimizing large problems unfortunately has overshadowed many simple problems in design practice which may be optimized relatively easily, such as the industrial applications reported in Chapter 3. The incorporation of simple optimization techniques in large programs to optimize a group of variables is quite common. While applictions of optimization in the auto industry could not be reported here, the existence of these applications is certain and quite encouraging.

6.1.1 Algorithm Efficiency

The efficiency of the structural optimization algorithm may be measured by: (1) the number of iterations; (2) the number of analyses of the structure; and (3) the total computational resources needed to obtain a minimum weight design. Most of the investigators use the first two criteria to present results. Each investigator, however, may have a different definition of what is exactly involved in an iteration or an analysis cycle. The last criterion, i.e., the total computational effort, is the most accurate, and of the total computational resources the computational effort on CPU time is the most important. The total CPU time required to optimize a structure may be divided into time spent in performing the different operations. The major operations are: (1) the analysis of the structure to determine the response of the structure to the applied loads; (2) the evaluation of the gradients of the objective and constraint functions; and (3) the modification of the design vector.

The largest segment of CPU time is usually spent on the analysis of the structure. This effort depends on the number of elements and the degrees of freedom of the nodes of the discretized structure. In order to reduce the computer time associated with an analysis, some investigators have proposed methods for rapid reanalysis of the structure.

The second most time consuming operation is the evaluation of the gradients of the constraints. The methods used to determine the gradients and the total number of constraints determine the CPU time spent on this operation. The methods used to evaluate the gradients of the constraints are based on using the pseudo load vectors or the unit load method. The second approach is generally found to be more efficient when only a few gradients are required.

The last operation is the modification of the design vector. This is normally achieved by using a recurrence relation where the incremental design vector is added to the old vector or the old design vector is multiplied by another vector. For some methods the recurrence relation contains the Lagrange multipliers as unknowns and one has to first evaluate them. The number of Lagrange multipliers corresponds to the number of constraints that are treated as potentially active. If this number is large the evaluation of the Lagrange multipliers needs more CPU time. The rate of percentage decrease in the weight of the structure depends on the recurrence relation selected to modify the design vector.

The decision as to which constraints are to be considered as active and passive affects the CPU time spent on the last two operations. Furthermore, approximations made in evaluating the gradients of the constraints also affect CPU time. Considering some important constraints as passive will give a near minimum weight design and not a minimum weight design. However, this approximation substantially reduces the total effort spent in the evaluation of the gradients and the Lagrange multipliers. A good example of this is when designing a structure with displacement and stress constraints, consider the stress constraints as passive constraints. When the number of design variables is of the order of a thousand and more, it is essential to make this approximation in order to solve the problem within a reasonable time and within the capability of the computer program. Even with a large number of displacement constraints, it is sometimes advantageous to pick the constraints at the most critical locations on the structure and design the structure on that basis.

6.1.2 Fully Stressed Design (FSD)

For the stress constraint problem the concept of FSD and uniform strain energy density are often used. When the allowable stress is the same for all elements, for a structure with all bar elements both these approaches are equivalent and give the same design; but for plate and beam elements, these two approaches will give a different distribution of material and different distribution of material and different designs. In a structure where there are small differences in the maximum allowable stress for the different elements and the degree of indeterminacy is low, the FSD algorithm gives a minimum weight design. However, for highly unequal maximum allowble stresses amongst different elements, the FSD algorithm will not only give a nonoptimum design but will also give a design with an inefficient load path and an improper distribution of the material in the structure. In spite of its inadequacies but because of its simplicity one sees from the literature reviewed in Section 4.2.1, that the FSD concept is often used to design aerospace structures. The main reason for the inadequacy of the FSD algorithm is that it makes the simplifying assumption that the gradient of the stress constraint of an element is independent of the other elements. For the case of a structure that is idealized with bar elements, where the allowable stress in a bar is specified by one value, the stress constraints can easily be included in the possible active constraint equations and a more rigorous optimality criterion approach can be used. In the case of plate elements, however, the strength criteria is a function of more than one stress and the problem of evaluating the stress gradients becomes difficult and time consuming. This is particularly true for fiber reinforced composite structures, where there are more than one layer in each element, and each layer has to satisfy a maximum stress or maximum strain or some other empirical strength criteria. In spite of these difficulties, in the design of a realistic structure it may be necessary to use a more rigorous mathematical programming or optimality criterion method rather than the simple FSD concept in order to obtain a proper distribution of the material and an efficient load path in the optimized structure.

6.2 DESIGN APPLICATIONS

Chapter 3 presented a variety of industrial applications of structural optimization. It is instructive to consider those cases where the methods have been implemented. In those situations where designers have an incentive to optimize it is still often possible to accomplish this goal with relatively straight-forward techniques and not mathematical programming. For example, most frames or trusses can come close to optimum by fully stressed iterative design, that is by producing a feasible design, reanalyze for forces then redesign, etc. This most often converges to a very acceptable design. Furthermore, experienced designers can approach an optimum by simple rules or judgments based on previous work. Some of the unique characteristics of successful applications of computer oriented direct optimization methods may be summarized as follows: (a) a large number of separate designs must be prepared as for design charts or sales tables. A computer is required for such extensive calculations and the variety of cases make single design rule inoperative; (b) highly competitive markets, such as standardized buildings, in which structural costs can affect the likelihood of a successful sale; (c) frequent changes to alternate material parameters or methods of fabrication may make historic criteria for choosing an optimum obsolete. It then becomes desirable to computerize the search for designs which are certain to be optima; (d) structures with tedious calculation procedures such as continuous span non-prismatic highway bridge girders in which a traditional analysis design cycle would require significant amounts of repetitive data preparation; and (e) unique applications in which structure weight must be minimized because of foundation or erection limitations.

In case where the strong rationalizations cited above are absent it is clear that most structural engineers will not use optimization techniques with their present limitations.

The incentive to optimize structures and proportion details in daily design practice is usually lacking. The explanation for this must be fully explored. In some cases an engineer's time is better spent examining a range of diverse structural solutions instead of refining a single one and this is certainly logical. In most cases the economic realities of the consulting practice do not permit sophisticated computer programs to be utilized for detailed proportioning. In fact if consulting fees are based on a percent of structure cost then savings in the latter actually reduce the consulting fee. More to the point may be the pressures to produce a set of drawings in a short period of time. While structural analysis programs have been available for some time (and not always used) these appear to most structural engineers as easy to use and applicable to a wide variety of cases. In constrast, optimization programs contain concepts and techniques not readily understood, they often require some user control and sophistication and a given program may only cover a single design application.

6.3 PROJECTIONS AND FUTURE NEEDS

6.3.1 Research

Progress is much needed in the areas of optimum geometry and topology. Optimum frame design is still considered complex and costly. Optimization of large systems and comparison with actual design continue to be of interest. Similarities between the iterative nonlinear analysis process and optimization should be exploited to reduce cost. Approximation methods are needed for checking and guiding the optimization process. In addition the iteraction and communication between researchers of numerical and analytical methods should be improved.

The effects of multiple loading on the optimum structure should be evaluated from the aspect of resistance to conditions which are not conveniently expressed in the problems constraints. Examples of such conditions are fatigue, damage tolerance, production sensitivity, and life cycle. It should, however, be pointed out that this same argument is frequently true for conventional design as well.

6.3.2 Applications

In the period immediately following its inception, structural optimization was expected to automate the design process and reduce the need for experienced designers. These unrealistically high expectations brought considerable disappointment to many professionals and designers, who could not justify the high costs of numerous analyses which led to highly theoretical results which, in turn, could not be readily applied to actual design problems. The outcome was a period of rejection and little progress in applications, while a backlog of theoretical development accumulated.

It is interesting to compare finite element methods and structural optimization, both developed during the past 20 years. Finite element methods enjoyed considerably more success than structural optimization because the theoretical foundation of the former was relatively well defined. Indeed finite element methods are actually applications of the theory of elasticity. Research in this case concentrated on the computational aspects. By constrast, structural optimization adopted mathematical programming techniques, which were not directly applicable to design problems. Hence, both theory and applications had to be developed.

It is becoming increasingly evident that progress in applying optimization theory to the design process will come through the programming of black boxes by optimization consultants and through their introduction to the design office. These black boxes must be programmed to comply with the special requirements and pecularities of the specific design applications, for which they are written. This process will, of course, involve a detailed study of the analysis programs, the fragmented design routines and the traffic of design data which exist in the office. This study should indicate where in a general integrated program should the optimization model reside and what its relation to other models is. On a smaller scale the advent of minicomputers and microprocessors will probably give the incentive for optimizing even specific standalone design problems. This situation will prevail until a sufficient number of engineers will acquire the basics of optimization. At that point the black boxes will be transparent in a similar process to that of finite element development. At their ultimate development, finite element and optimization methods will be a natural part of design and analysis.

A practical approach to design, currently being developed, combines SLP (sequential linear programming) methods and FEM (finite element methods). The advantage of this approach is that both SLP and FEM are well defined and rather well understood by many engineers. Convergence using these methods is usually reached in as few as approximately 10 iterations.

The necessary ingredients for the process described above are in the introduction of structural optimization concepts in engineering curriculum and in continuing education programs for practicing engineers.

Techniques for data management must be developed to facilitate the incorporation of optimization model in existing analysis or design programs.

In any case, structural optimization has been firmly established as an important tool for research and development of structural design. At present several important factors will contribute to facilitate, justify, and enhance its further development. These factors are: the decreasing cost of hardware, the increasing technological developments, the energy crisis, and the necessity to improve productivity.

7. FREQUENTLY-REFERRED-TO-EXAMPLES

List of Figures

Figure		Page
1.	Nine Bar Plane Truss	71
2.	Eleven Bar Truss	72
3.	Ten Bar Cantilevered Truss	73
4.	Two Hundred Member Cantilevered Truss	74
5.	Nine Bar Space Truss	75
6.	Twenty-five Member Space Tower	76
7.	Seventy-two Bar Space Tower	78
8.	Conventional Transmission Tower	80
9.	Forty-seven Member Transmission Tower	81
10.	Portal Frame	82
11.	One Bay - Two Story Frame	83
12.	One Bay - Two Story Frame Rigid Truss Frame	85
13.	Two Bay - Two Story Frame	86
14.	Two Bay - Six Story Rigid Frame	87
15.	Two Member Grid Structure	88
16.	Six Member Grid Structure	89
17.	Delta Wing Structure	90
18.	Elevated Transit Structure	91
19.	Welded Plate Girder	92

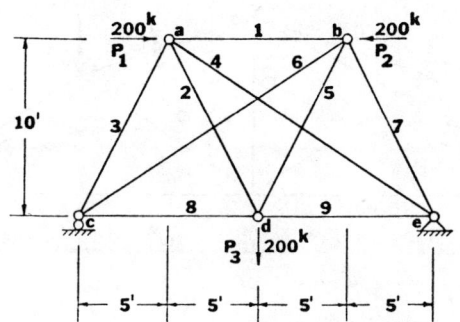

Figure 1. Nine Bar Plane Truss

1. Farshi, B. and Schmit, L. A., "Minimum Weight of Stress Limited Trusses," *Journal of the Structural Division*, ASCE, Vol. 100, ST1, January 1974.

2. Reinschmidt, K., "Discrete Structural Optimization," *Journal of the Structural Division*, ASCE, Vol. 97, ST1, January 1971.

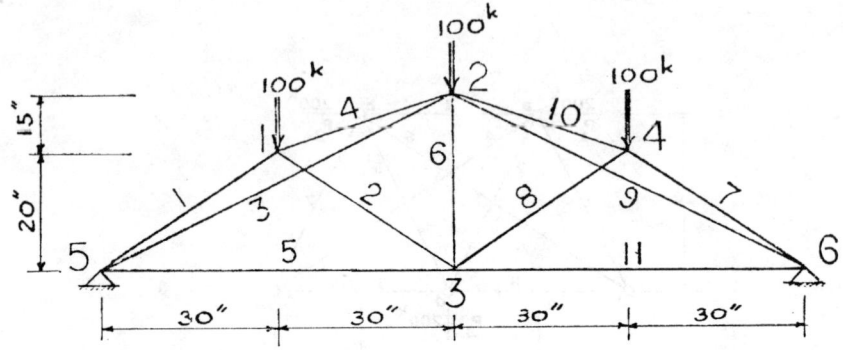

Figure 2. Eleven Bar Truss

1. Lipson, S. L. and Agrawal, N. M., "Weight Optimization of Plant Trusses," <u>Journal of the Structural Division</u>, ASCE, Vol. 100, No. ST5, Proc. Paper 10521, May 1974, pp. 765-879.

2. Saka, M. P., "Shape Optimization of Trusses," <u>Journal of the Structural Division</u>, ASCE, Vol. 106, No. ST5, Proc. Paper 15437, May 1980, pp. 115-1174.

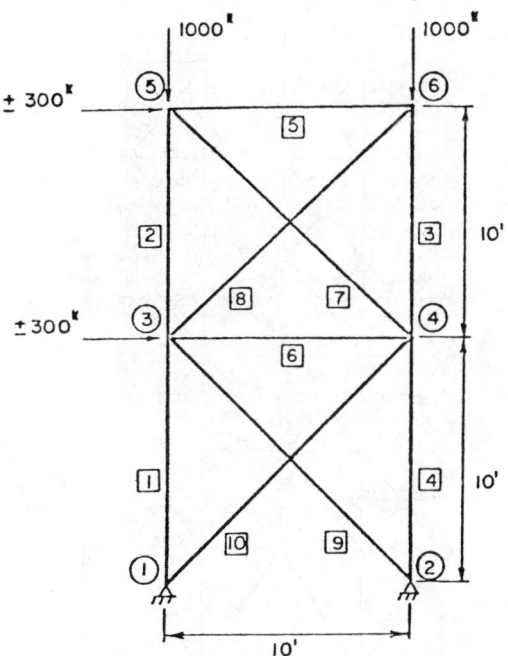

Figure 3. Ten Bar Cantilevered Truss

1. Arora, J. S. and Haug, E. J., "Efficient Hybrid Methods of Optimal Structural Design," *Journal of the Engineering Mechanics Division*, ASCE, Vol. 104, No. EM3, June 1978.

2. Farshi, B. and Schmit, L. A., "Minimum Weight of Stress Limited Trusses," *Journal of the Structural Division*, ASCE, Vol. 100, ST1, January 1974.

3. Felton, L. P. and Dobbs, M. W., "On Optimized Prestressed Trusses," *AIAA Journal*, Vol. 15, No. 7, July 1977, pp. 1037-1039.

4. Gellatly, R. A. and Berke, L., "Optimal Structural Design," AFFDL-TR-70-165.

5. Sheu, C. Y. and Schmit, L. A., "Minimum Weight Design of Elastic Redundant Trusses Under Multiple Static Loading Conditions," *AIAA Journal*, Vol. 10, No. 2, Feburary 1972.

6. Venkayya, V. B., "Design of Optimum Structures," *Computers and Structures*, Vol. 1, No. 1/2, 1971, pp. 265-309.

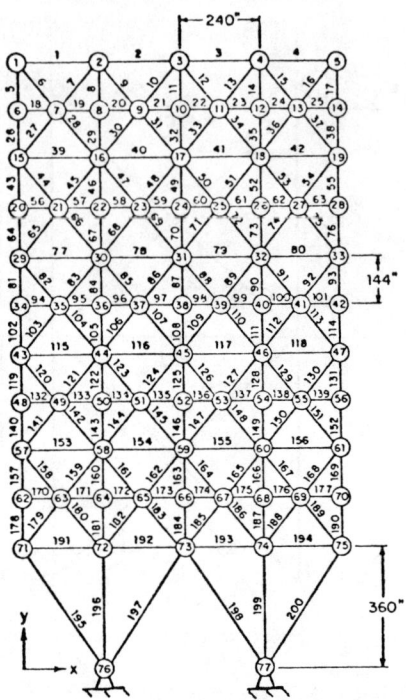

Figure 4. Two Hundred Member Cantilevered Truss

1. Arora, J. S. and Govil, A. K., "An Efficient Method for Optimal Structural Design by Substructuring," Int. J. of Computers and Structures, Vol. 7, No. 4, 1977.

2. Arora, J. S. and Govil, A. K., "Design Sensitivity Analysis with Substructuring," Journal of the Engineering Mechanics Division, ASCE, Vol. 103, No. EM4, August 1977.

3. Arora, J. S. and Haug, E. J., Jr., "Efficient Optimal Design of Structures by Generalized Steepest Descent Programming," Int. J. for Numerical Methods in Engineering, Vol. 10, 1976, pp. 747-766.

4. Pickett, R. M., Jr., Rubenstein, M. F., and Nelson, R. B., "Automated Structural Synthesis Using a Reduced Number of Design Coordinates," AIAA Journal, Vol. 11, No. 4, April 1973, pp. 489-494.

5. Venkayya, V. B., Khot, N. S., and Reddy, V. S., "Optimization of Structures Based on the Study of Energy Distribution," AFFDL-TR-68-150, 1969.

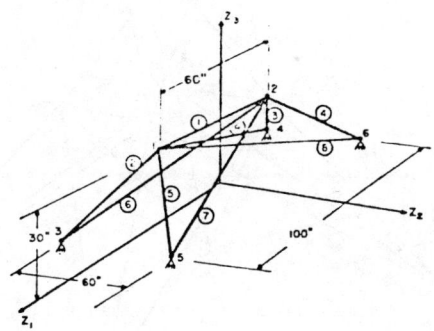

Figure 5. Nine Bar Space Truss

1. Felton, L. P. and Hoffmeister, L. D., "Optimized Components in Truss Synthesis," AIAA Journal, Vol. 6, No. 12, December 1968.

2. Felton, L. P., Nelson, R. B., and Bronowicki, A. V., "Thin Walled Elements in Truss Synthesis," AIAA Journal, Vol. 11, No. 12, December 1973, pp. 1780-1782.

3. Fox, R. L. and Schmit, L. A., "Advances in the Integrated Approach to Structural Synthesis," Journal of Spacecraft and Rockets, Vol. 3, No. 6, June 1966.

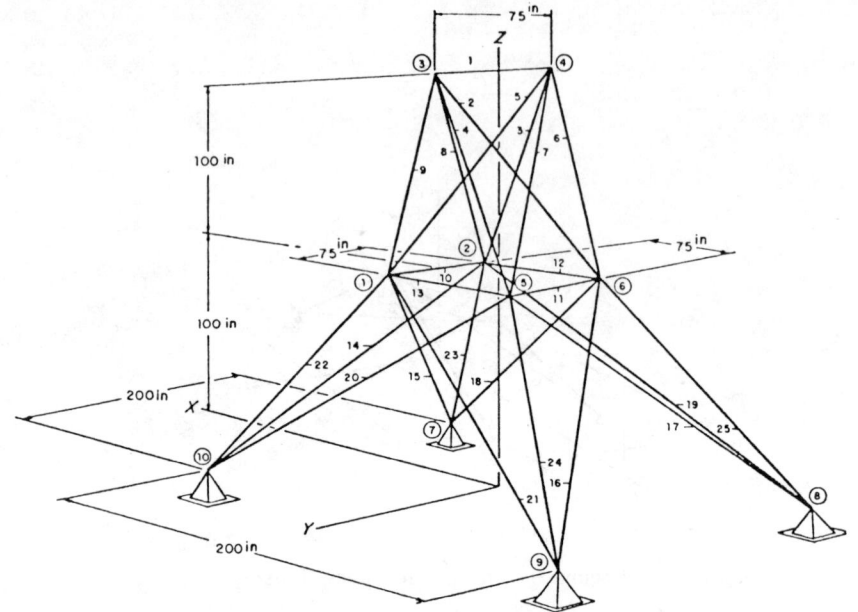

Figure 6. Twenty-five Member Space Tower

1. Arora, J. S. and Govil, A. K., "An Efficient Method for Optimal Structural Design by Substructuring," <u>International Journal of Computers and Structures</u>, Vol. 7, No. 4, 1977, pp. 507-515.

2. Arora, J. S. and Haug, E. J., "Efficient Optimal Design of Structures by Generalized Steepest Descent Programming," <u>International Journal for Numerical Methods in Engineering</u>, Vol. 10, 1976, pp. 1420-1427.

3. Dobbs, M. W. and Nelson, R. B., "Application of Optimality Criteria to Automated Structural Design," <u>AIAA Journal</u>, Vol. 14, No. 10, October 1976, pp. 1436-1443.

4. Dwyer, W. J., Emerton, R. K., and Sabatell, P. L., "An Automated Procedure for the Optimization of Practical Aerospace Structures," AFFDL-TR-70-118.

5. Felton, L. P. and Dobbs, M. W., "On Optimized Prestressed Trusses," <u>AIAA Journal</u>, Vol. 15, No. 7, July 1977, pp. 1037-1039.

6. Fox, R. L. and Schmidt, L. A., "Advances in the Integrated Approach to Structural Synthesis," <u>Journal of Spacecraft and Rockets</u>, Vol. 3, No. 6, June 1966.

7. Gellatly, R. A. and Berke, L., "Optimal Structural Design," AFFDL-TR-70-165.

8. Gellatly, R. A. and Gallagher, R. H., Development of Advanced Structural Optimization Programs and Their Application to Large Order Systems, AFFDL-TR-66-80, Flight Dynamics Laboratory, Wright-Patterson AFB, 1965.

9. Pradad, B. and Haftka, R. T., "A Cubic Extended Interior Penalty Function for Structural Optimization," Int. J. for Numerical Methods in Engineering, Vol. 14, No. 8, 1979, pp. 1107-1126.

10. Schmit, L. A. and Miura, H., "Approximation Concepts for Efficient Structural Synthesis," NASA CR-2552, 1975.

11. Schmit, L. A., Jr. and Miura, H., "A New Structural Analysis/Synthesis Capability - Access 1," AIAA Paper 75-763, 1975.

12. Vanderplaats, G. N. and Moses, F., "Automated Design of Trusses for Optimum Geometry," Journal of the Structural Division, ASCE, Vol. 98, ST3, March 1972.

13. Vanderplaats, G. N. and Moses, F., "Structural Optimization by Methods of Feasible Directions," Int. J. of Computers and Structures, Vol. 3, 1973, pp. 739-755.

14. Venkayya, V. B., "Design of Optimum Structures," Computers and Structures, Vol. 1, No. 1/2, August 1971, pp. 265-309.

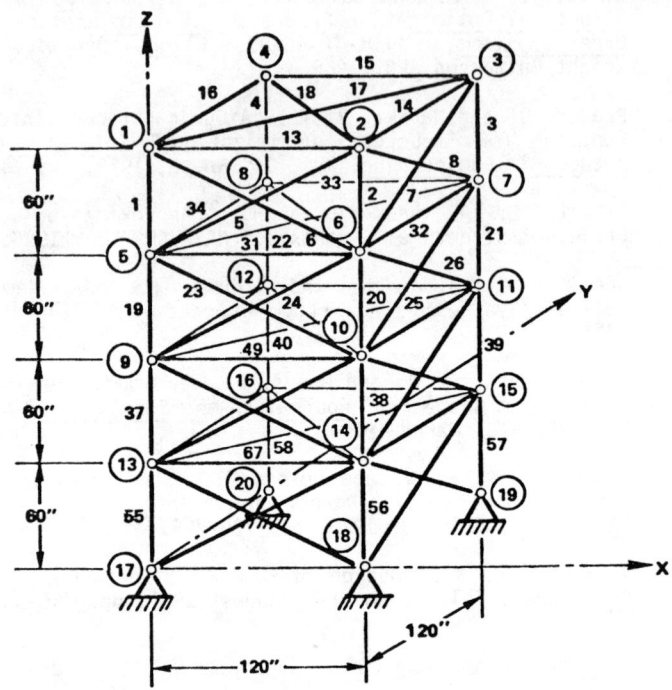

Figure 7. Seventy-two Bar Space Tower

1. Arora, J. S. and Haug, E. J., Jr., "Efficient Optimal Design of Structures by Generalized Steepest Descent Programming," Int. J. for Numerical Methods in Engineering, Vol. 10, 1976, pp. 747-766.

2. Dwyer, W. J., Emerton, R. K., and Sabatelli, P. L., "An Automated Procedure for the Optimization of Practical Aerospace Structures," AFFDL-TR-70-118.

3. Felton, L. P. and Dobbs, M. W., "On Optimized Prestressed Trusses," AIAA Journal, Vol. 15, No. 7, July 1977, pp. 1037-1039.

4. Gellatly, R. A. and Berke, L., "Optimal Structural Design," AFFDL-TR-70-165, 1971.

5. Haftka, R. T. and Prasad, B., "Programs for Analysis and Resizing of Complex Structures," Int. J. of Computers and Structures, Vol. 10, No. 1/2, 1979, pp. 323-330.

6. Fleury, C. and Schmit, L. A., "Primal and Dual Methods in Structural Optimization," Journal of the Structural Division, ASCE, Vol. 106, No. ST5, Proc. Paper 15431, May 1980, pp. 1117-1133.

7. Khot, N. S., Venkayya, V. B., and Berke, L., "Optimization of Structures for Strength and Stability Requirements," AFFDL-TR-73-98.

8. Schmit, L. A. and Farshi, B., "Some Approximation Concepts for Structural Synthesis," <u>AIAA Journal</u>, Vol. 12, No. 5, May 1974.

9. Venkayya, V. B., "Design of Optimum Structures," <u>Int. J. of Computers and Structures</u>, Vol. 1, No. 1/2, 1971, pp. 265-309.

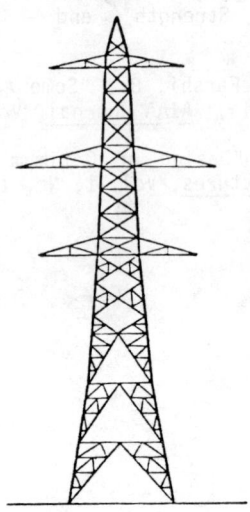

Figure 8. Conventional Transmission Tower

1. Raj, P. P. and Durrant, S. O., "Optimum Structural Design by Dynamic Programming," <u>Journal of the Structural Division</u>, ASCE, Vol. 102, ST8, August 1976.

2. Sheppard, D. J. and Palmer, A. C., "Optimal Design of Transmission Towers by Dynamic Programming," <u>Computers and Structures</u>, Vol. 2, No. 4, 1972, pp. 455-468.

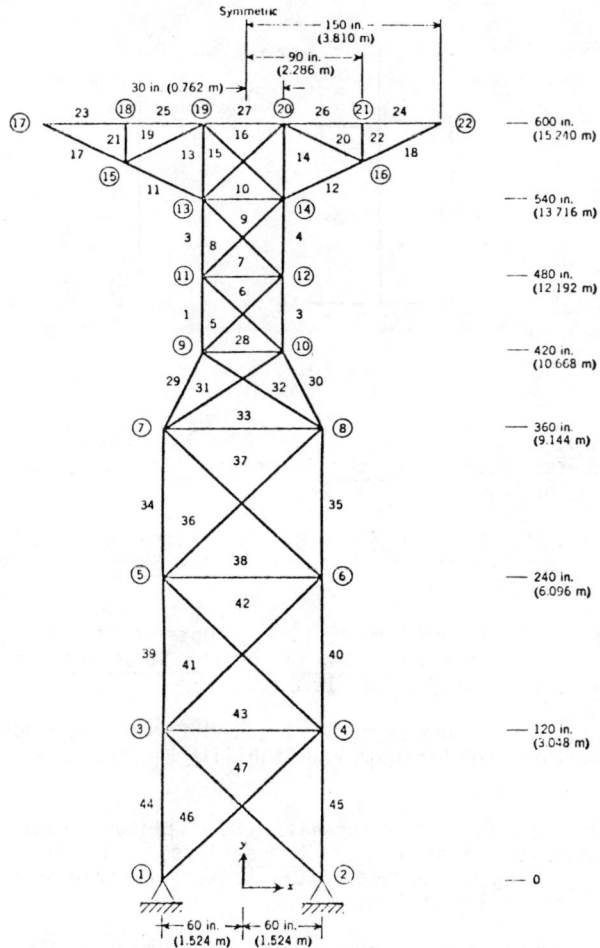

Figure 9. Forty-seven Member Transmission Tower

1. Lipson, S. L. and Agrawal, K. M., "Weight Optimization of Plane Trusses," <u>Journal of the Structural Division</u>, ASCE, Vol. 100, ST5, May 1974.

2. Twisdale, L. A. and Khachaturian, N., "Multistage Optimization of Structures," <u>Journal of the Structural Division</u>, ASCE, Vol. 101, ST5, May 1975.

3. Vanderplaats, G. N. and Moses, F., "Automated Design of Trusses for Optimum Geometry," <u>Journal of the Structural Division</u>, ASCE, Vol. 98, ST3, March 1972.

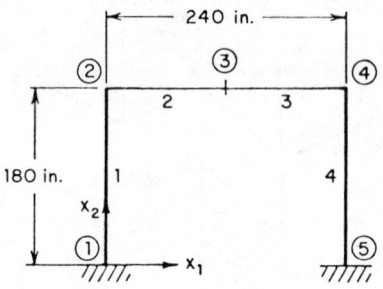

Figure 10. Portal Frame

1. Arora, J. S., Haug, E. J., and Rim, K., "Optimal Design of Plane Frames," <u>Journal of the Structural Division</u>, ASCE, Vol. 101, ST10, October 1975.

2. Brown, D. M. and Ang, H. S., "Structural Optimization by Nonlinear Programming," <u>Journal of the Structural Division</u>, ASCE, Vol. 92, ST6, December 1966.

3. Cassis, J. H. and Schmit, L. A., "Optimum Structural Design with Dynamic Constraints," <u>Journal of the Structural Division</u>, ASCE, Vol. 102, ST10, October 1976.

4. Khot, N. S., Venkayya, V. B., and Berke, L., "Optimization of Structures for Strength and Stability Requirements," AFFDL-TR-73-98.

5. Kaun-Chen Fu and Kuang-Wei You, "Optimum Frame Design Using Available Sections," <u>Proceedings</u>, National Structural Engineering Conference, ASCE, University of Wisconsin, Madison, Wisconsin, August 1976.

6. Kuzmanovic, B. O. and Willems, N., "Optimum Plastic Design of Steel Frames," <u>Journal of the Structural Division</u>, ASCE, Vol. 98, ST8, August 1972.

7. Shamie, J. and Schmit, L. A., "Frame Optimization Including Frequency Constraints," <u>Journal of the Structural Division</u>, ASCE, Vol. 101, ST1, January 1975.

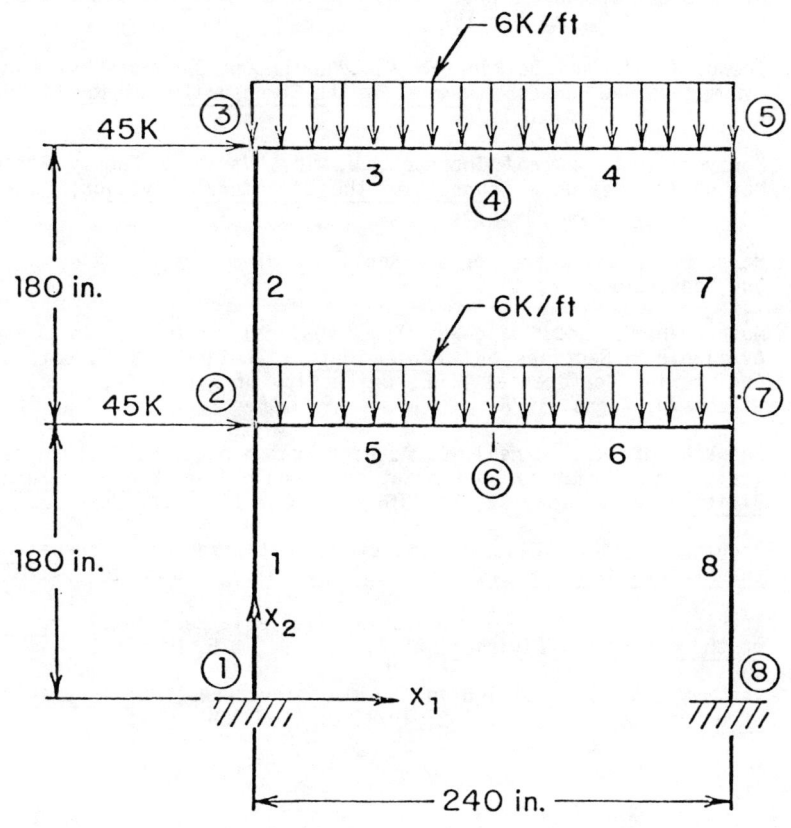

Figure 11. One Bay - Two Story Frame

1. Arora, J. S. and Govil, A. K., "An Efficient Method for Optimal Structural Design by Substructuring," *Int. J. of Computers and Structures*, Vol. 7, No. 4, 1977, pp. 507-515.

2. Arora, J. S., Haug, E. J., and Rim, K., "Optimal Design of Plane Frames," *Journal of the Structural Division*, ASCE, Vol. 101, ST10, October 1975, pp. 2963-3078.

3. Brown, D. M. and Ang, H. S., "Structural Optimization by Nonlinear Programming," *Journal of the Structural Division*, ASCE, Vol. 92, ST6, December 1966.

4. Cassis, J. H. and Schmit, L. A., "Optimum Structural Design with Dynamic Constraints," *Journal of the Structural Division*, ASCE, Vol. 102, ST10, October 1976.

5. Cheng, F. Y. and Botkin, M. E., "Nonlinear Optimum Design of Dynamic Damped Frames," *Journal of the Structural Division*, ASCE, Vol. 102, ST3, March 1976.

6. Gorzynski, J. W. and Thornton, W. A., "Variable Energy Ratio Method for Design," *Journal of the Structural Division*, ASCE, Vol. 101, ST4, April 1975.

7. Haug, E. J. and Arora, J. S., *Applied Optimal Design*, John Wiley and Sons, New York, 1979.

8. Kuan-Chen Fu and Kuang-Wei You, "Optimal Frame Design Using Available Sections," *Proceedings*, National Structural Engineering Conference, ASCE, University of Wisconsin, Madison, Wisconsin, August 1976.

9. Reinschmidt, K., Cornell, C. A., and Brotchie, J. F., "Iterative Design and Structural Optimization," *Journal of the Structural Division*, ASCE, Vol. 92, No. ST6, December 1966, pp. 281-318.

10. Reinschmidt, K., "Discrete Structural Optimization," *Journal of the Structural Division*, ASCE, Vol. 97, ST1, January 1971.

11. Toakley, R., "Optimum Design Using Available Sections," *Journal of the Structural Division*, ASCE, Vol. 94, ST5, May 1968.

12. Venkayya, V. B., "Design of Optimum Structures," *Computers and Structures*, Vol. 1, No. 1/2, 1971, pp. 265-309.

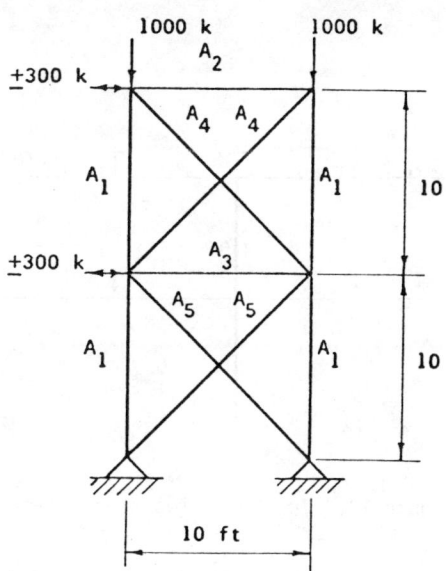

Figure 12. One Bay - Two Story Rigid Truss Frame

1. Gorzynski, J. W. and Thornton, W. A., "Variable Energy Ratio Method for Design," <u>Journal of the Structural Division</u>, ASCE, Vol. 101, ST4, April 1975.

2. Thomas, H. R. and Brown, D. M., "Optimum Least-Cost Design of a Truss Roof System," <u>Computers and Structures</u>, Vol. 6, No. 1, February 1977.

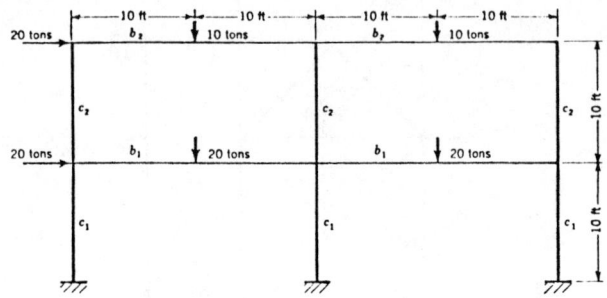

Figure 13. Two Bay - Two Story Frame

1. Cassis, J. H. and Schmidt, L. A., "Optimum Structural Design with Dynamic Constraints," Journal of the Structural Division, ASCE, Vol. 102, ST10, October 1976, pp. 2053-2071.

2. Cheng, F. Y. and Botkin, M. E., "Nonlinear Optimum Design of Dynamic Damped Frames," Journal of the Structural Division, ASCE, Vol. 102, ST3, March 1976.

3. Frind, E. O. and Wright, P. M., "Gradient Methods in Optimum Design," Journal of the Structural Division, ASCE, Vol. 101, ST4, April 1975.

4. Haug, E. J., Arora, J. S., and Feng, T. T., "Sensitivity Analysis and Optimization of Structures for Dynamic Response," Journal of Mechanical Design, ASME, Vol. 100, April 1978.

5. Kuan-Chen Fu and Kuang-Wei You, "Optimal Frame Design Using Available Sections," Proceedings, National Structural Engineering Conference, ASCE, University of Wisconsin, Madison, Wisconsin, August 1976.

6. Toakley, R., "Optimum Design Using Available Sections," Journal of the Structural Division, ASCE, Vol. 94, ST5, May 1968.

7. Venkayya, V. B., "Design of Optimum Structures," Computers and Structures, Vol. 1, No. 1/2, 1971, pp. 265-309.

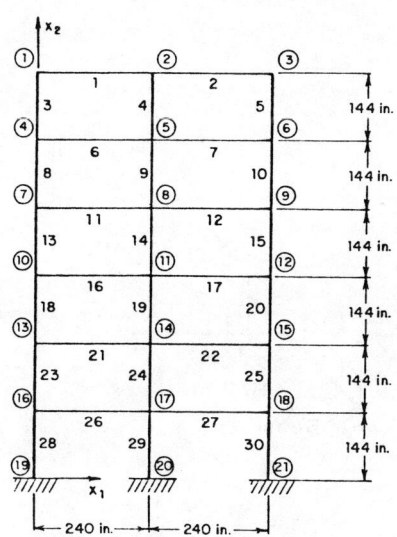

Figure 14. Two Bay - Six Story Rigid Frame

1. Arora, J. S., Haug, E. J., and Rim, K., "Optimal Design of Plane Frames," <u>Journal of the Structural Division</u>, ASCE, Vol. 101, ST10, October 1975, pp. 2963-3078.

2. Haug, E. J. and Arora, J. S., <u>Applied Optimal Design</u>, John Wiley and Sons, New York, 1979.

3. Reinschmidt, K., Cornell, C. A., and Brotchie, J. F., "Iterative Design and Structural Optimization," <u>Journal of the Structural Division</u>, ASCE, Vol. 92, No. ST6, December 1966, pp. 281-318.

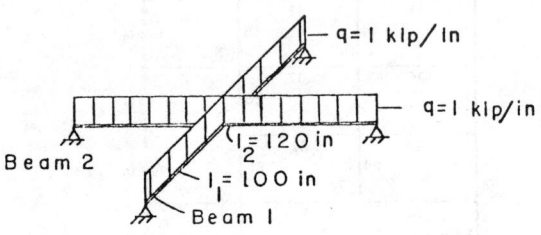

Figure 15. Two Member Grid Structure

1. Moses, F. and Onoda, S., "Minimum Weight Design of Structures with Application to Elastic Grillages," <u>International Journal for Numerical Methods in Engineering</u>, Vol. 1, 1969.

2. Reinschmidt, K. and Norabhoompipat, T., "Structural Optimization by Equilibrium Linear Programming," <u>Journal of the Structural Division</u>, ASCE, Vol. 101, ST4, April 1975.

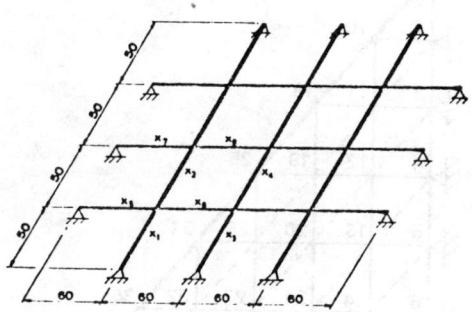

Figure 16. Six Member Grid Structure

1. Kavlie, D. and Moe, J., "Automated Design of Frame Structures," Journal of the Structural Division, ASCE, Vol. 97, ST1, January 1971.

2. Reinschmidt, K. and Norabhoompipat, T., "Structural Optimization by Equilibrium Linear Programming," Journal of the Structural Division, ASCE, Vol. 101, ST4, April 1975.

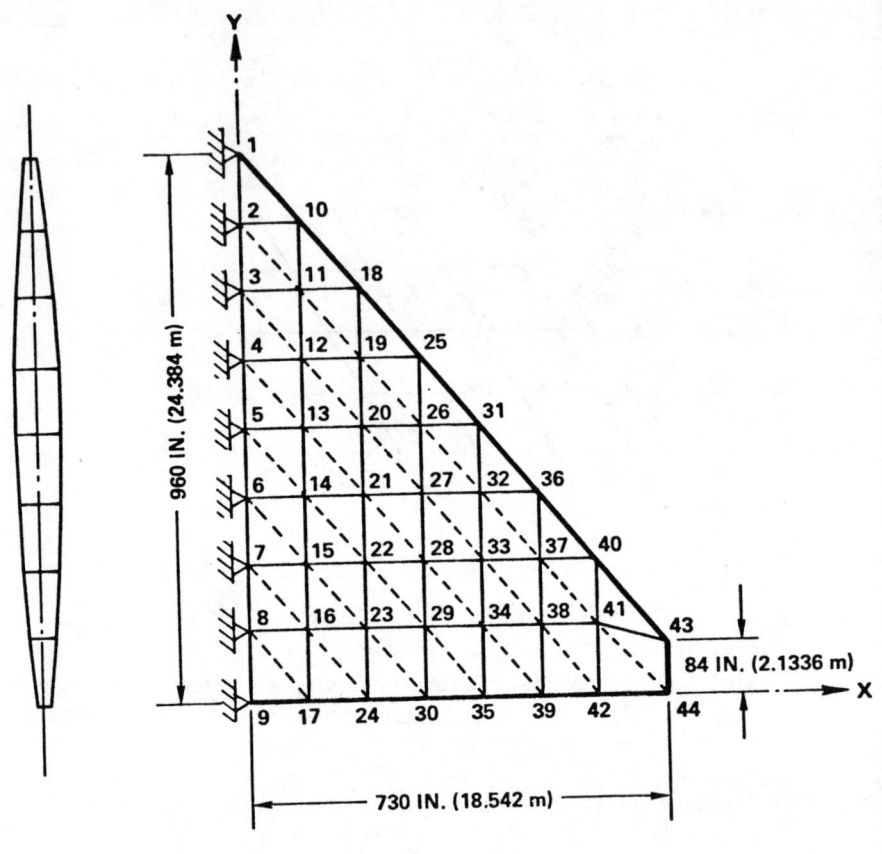

Figure 17. Delta Wing Structure

1. Haftka, R. T. and Starnes, J. H., Jr., "Application of a Quadratic Extended Interior Penalty Function for Structural Optimization," AIAA Journal, Vol. 14, No. 6, June 1976, pp. 718-724.

2. Schmit, L. A. and Miura, H., "Approximation Concepts for Efficient Structural Synthesis," NASA CR-2552, 1976.

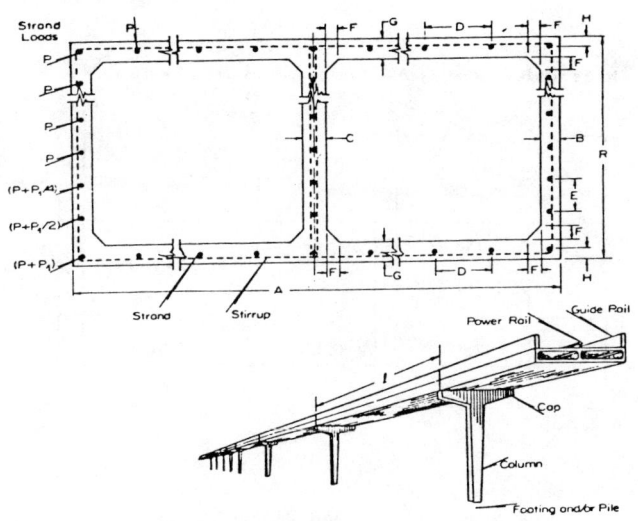

Figure 18. Elevated Transit Structure

1. Touma, A. and Wilson, J. F., "Design Optimization of Prestressed Concrete Spans for High Speed Ground Transportation," Int. J. of Computers and Structures, Vol. 3, No. 2, 1973, pp. 265-280.

2. Naaman, A. E. and Silver, M. L., "Minimum Cost Design of Elevated Transit Structures," Journal of the Construction Division, ASCE, Vol. 102, CO1, March 1976.

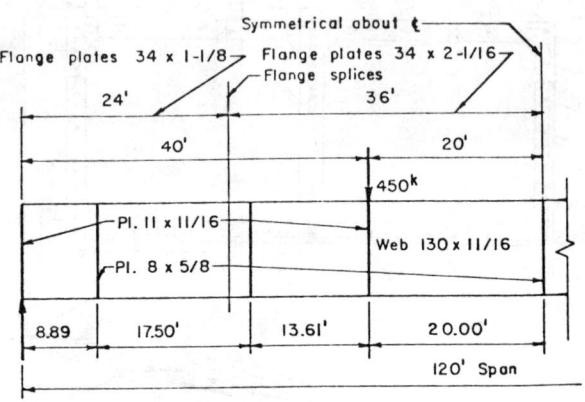

Figure 19. Welded Plate Girder

1. Annamalai, N., Lewis, A. D. M., and Goldberg, J. E., "Cost Optimization of Welded Plate Girders," <u>Journal of the Structural Division</u>, ASCE, Vol. 98, ST10, October 1972.

2. Goble, G. and DeSantis, "Optimum Design of Mixed Steel Composite Girders," <u>Journal of the Structural Division</u>, ASCE, Vol. 92, ST6, December 1966, pp. 24-44.

8. REFERENCES

8.1 BOOKS ON STRUCTURAL OPTIMIZATION

1. Cohn, M. Z. (Editor), *An Introduction to Structural Optimization*, Solid Mechanics Div. Study No. 1, Univ. of Waterloo, 1969.

2. Cox, H. L., *The Design of Structures of Least Weight*, Pergamon Press, 1965.

3. Cyras, A. A., Brokouskas, A. E., and Karkauskas, R. P., *Theory and Methods of Optimization of Elastic-Plastic Systems*, (in Russian), Strojizdat, 1974.

4. Fox, R. 1., *Optimization Methods For Engineering Design*, Addison-Wesley, 1971.

5. Gallagher, R. H. and Zienkiewicz (editors), *Optimal Structural Design, Theory and Applications*, (Proc. International Symp. on Optimization of Structural Design, Swansea 1972), John Wiley & Sons, New York, 1973.

6. Gerard, G., *Minimum-Weight Design of Compression Structures*, University Press, New York, 1956.

7. Griniev, V. B. and Philipov, A. P., *Optimization of Structural Elements for Mechanical Constraints*, (in Russian) Naukova Dumka, Kiev, 1975.

8. Haug, E. J. and Arora, J. S., *Applied Optimal Design*, Wiley-Interscience, New York, 1979.

9. Hemp, W. S., *Optimum Structures*, Clarendon Press, Oxford, 1973.

10. Kirsch, U., *Optimum Structural Design: Concepts, Methods and Applications*, McGraw Hill, New york, 1981.

11. Majid, K. I., *Optimum Design of Structures*, Neunes-Butterworths, 1974.

12. Prager, W., *Introduction to Structural Optimization*, Courses and Lectures: International Centre for Mechanical Science, Vdine, No. 212, Springer-Verlag, Vienna, 1974.

13. Reitman, M. I. and Shapiro, G. S., *Methods of Optimal Design of Deformable Bodies*, (in Russian) Nauku, Moscow, 1976.

14. Rozvany, G. I. N., *Optimal Design of Flexural Systems*, Pergamon Press, London, 1976.

15. Sawcyuk, A. and Mroz, A. (editors), *Optimization in Structural Design*, (Proc. IUTAM Symp. Warsaw, 1973) Springer-Verlag, Berlin, 1975.

16. Shanley, F. R., *Weight/Strength Analysis of Aircraft Structures*, Dover, New York, 1960.

17. Spillers, W. R., *Iterative Structural Design*, North Holland, Amsterdam, 1975.

18. Spunt, L., *Optimum Structural Design*, Prentice Hall, Englewood Cliffs, New Jersey, 1971.

19. Troickij, V. A., *Optimal Processes of Vibrations in Mechanical Systems*, (in Russian) Mashinostrojenije, Leningrad, 1976.

8.2 GENERAL BOOKS ON OPTIMIZATION

1. Abadie, J. (Editor), <u>Integer and Nonlinear Programming</u>, North Holland Publishing Company, Amsterdam, 1970.
2. Abadie, J. (Editor), <u>Nonlinear Programming</u>, North Holland Publishing Company, Amsterdam, 1967.
3. Adby, P. R. and Dempster, M. A., <u>Introduction to Optimization Methods</u>, Halsted Press, 1974.
4. Anderson, R. S., et al., eds. <u>Optimization</u>, U. of Queensland Press, 1972.
5. Aoki, M., <u>Optimization of Stochastic System</u>, The MacMillan Company, New York, 1971.
6. Aoki, M., <u>Optimization of Stochastic Systems</u>, Vol. 32, Mathematics in Science and Engineering, Academic Press, New York, 1967.
7. Arrow, K., Hurwicz, L., and Uzawa, H., (Editors), <u>Studies in Linear and Nonlinear Programming</u>, Stanford University Press, 1958.
8. Avriel, Mordecai, <u>Nonlinear Programming</u>, Prentice Hall, 1976.
9. Balakrishnan, A. V., <u>Introduction to Optimization Theory in a Hilbert Space</u>, Springer-Verlag, 1971.
10. Balinski, M. L. and Wolfe, P., <u>Nondifferentiable Optimization</u>, Elsevier, 1976.
11. Bazaraa, M. S. and Shetty, C. M., <u>Foundations of Optimization</u>, Springer-Verlag, 1976.
12. Beale, E. M. L., (Editor), <u>Applications of Mathematical Programming Techniques</u>, American Elsevier Publishing Company, Inc., New York, 1970.
13. Beale, E. M. L., <u>Mathematical Programming in Practice</u>, John Wiley and Sons, Inc., New York, 1968.
14. Beightler, Charles S. and Phillips, Donald T., <u>Applied Geometric Programming</u>, Wiley, 1976.
15. Bellman, R. (Editor), <u>Mathematical Optimization Techniques</u>, University of California Press, Berkeley and Los Angeles, 1963.
16. Bellman, R. E., <u>Dynamic Programming</u>, Princeton University Press, 1957.
17. Bellman, R. E. and Dreyfus, S., <u>Applied Dynamic Programming</u>, Princeton University Press, 1962.

18. Beltrami, E. J., *Algorithmic Approach to Nonlinear Analysis and Optimization*, Academic Press, 1970.

19. Beveridge, Gordon S. and Schechter, Robert S., *Optimization Theory and Practice*, McGraw, 1970.

20. Bracken, J. and McCormick, G. P., *Selected Applications of Nonlinear Programming*, John Wiley and Sons, Inc., New York, 1968.

21. Burley, D. M., *Studies in Optimization*, Halsted Press, 1974.

22. Charnes, A. and Cooper, W. W., *Management Models and Industrial Applications of Linear Programming*, Vol. 1, John Wiley and Sons, Inc., New York, 1961.

23. Claycombe, William W. and Sullivan, Wilham G., *Foundations of Mathematical Programming*, Reston, 1975.

24. Collatz, L. and Wetterling, W., *Optimization Problems*, Springer-Verlag, 1975.

25. Converse, A. O., *Optimization*, Krieger, 1970.

26. Cooper, Leon and Steinberg, David, *Introduction to Methods of Optimization*, Saunders, 1970.

27. Dano, S., *Linear Programming in Industry*, Third Edition, Springer-Verlag, New York, 1965.

28. Dantzig, G. B., *Linear Programming and Extensions*, Princeton University Press, 1963.

29. Denn, Morton M., *Optimization by Variational Methods*, Krieger, 1978.

30. De Veubeke, B. F., *Advanced Problems and Methods of Space Flight Optimization*, Pergamon, 1969.

31. Dixon, L. C. W., *Nonlinear Optimization*, Crane, Russak and Company, Inc., New York, 1972.

32. Dorny, C. Nelson, *A Vector Space Approach to Models and Optimization*, Wiley, 1975.

33. Duffin, Richard J., et al., *Geometric Programming: Theory and Application*, Wiley, 1967.

34. Fiacco, A. V. and McCormick, G. P., *Nonlinear Programming: Sequential Unconstrained Minimization Techniques*, John Wiley and Sons, Inc., New York, 1968.

35. Fletcher, R. (Editor), *Optimization*, Academic Press, New York, 1969.

36. Fox, R. L., *Optimization Methods for Engineering Design*, Addison-Wesley Publishing Company, Reading, Massachusetts, 1971.

37. Garfinkel, R. S. and Nemhauser, G. L., *Integer Programming*, John Wiley and Sons, Inc., New York, 1972.

38. Garvin, W. W., *Introduction to Linear Programming*, McGraw-Hill Book Company, Inc., New York, 1960.

39. Gass, S. I., *Linear Programming*, Third Edition, McGraw-Hill Book Company, Inc., New York, 1969.

40. Geoffrion, A. M. (Editor), *Perspectives on Optimization: A Collection of Expository Articles*, Addison-Wesley Publishing Company, Reading Massachusetts, 1972.

41. Gill, P. E. and Murray, W., *Numerical Methods for Constrained Optimization*, Academic Press, 1974.

42. Girsanov, L. V., *Lectures on Mathematical Theory of Extremum Problems*, Springer-Verlag, 1972.

43. Gottfried, Bryon S. and Weisman, Joel, *Introduction to Optimization Theory*, Prentice-Hall, 1973.

44. Graves, R. L. and Wolfe, W., (Editors), *Recent Advances in Mathematical Programming*, McGraw-Hill Book Company, Inc., New York, 1963.

45. Greenberg, H., "Interger Programming", Vol. 76, *Mathematics in Science and Engineering*, Academic Press, New York, 1971.

46. Greenwald, D. U., *Linear Programming - An Explanation of the Simplex Algorithm*, The Ronald Press Company, New York, 1957.

47. Hadley, G., *Linear Programming*, Addison-Wesley Publishing Company, Reading, Massachusetts, 1962.

48. Hadley, G., *Nonlinear and Dynamics Programming*, Addison-Wesley, 1964.

49. Hammer, P. L. and Zoutendijk, G., *Mathematical Programming in Theory and Practice*, Elsevier, 1975.

50. Hasdorff, Lawrence, *Gradient Optimization and Nonlinear Control*, Wiley, 1976.

51. Hestenes, Magnus, R., *Optimization Theory: The Finite Dimensional Case*, Wiley, 1975.

52. Hillier and Lieberman, *Operations Research*, 2nd Ed., Holden Day, 1974.

53. Himmelblau, David, *Applied Nonlinear Programming*, McGraw Hill, 1972.

54. Holmes, R. B., *A Course on Optimization and Best Approximation*, Springer-Verlag, 1972.

55. Hu, T. C., *Integer Programming and Network Flows*, Addison-Wesley Publishing Company, Reading, Massachusetts, 1969.

56. Jacobs, O. L. R., *An Introduction to Dynamic Programming*, Chapman and Hall, Ltd., London, 1967.

57. Kaufmann, A. and Cruon, R., *Dynamic Programming, Sequential Scientific Management*, translated by Sneyd, H. C., Vol. 37, Mathematics in Science and Engineering, Academic Press, New York, 1967. (Original French Edition: "La Programmation Dynamique; Gestion Scientifique Sequentielle", Dunod, Paris, 1965).

58. Koo, D. D., *Elements of Optimization*, Springer-Verlag, 1977.

59. Kowalik, J. and Osborne, M. R., *Methods for Unconstrained Optimization Problems*, Elsevier, 1969.

60. Kuenzi, H. P., Krelle, W., and Oettli, W., *Nonlinear Programming*, Blaisdell Publishing Company, New York, 1966.

61. Kuhn, H. W. and Tucker, A. W., (Editors), *Linear Inequalities and Related Systems*, Annals of Mathematics Studies No. 38, Princeton University Press, 1956.

62. Kuenzi, Hans P., et al., *Numerical Methods of Mathematical Optimization with ALGOL and FORTRAN Programs*, Academic Press, 1968.

63. Larson, Robert E., *State Space Increment Dynamic Programming*, Elsevier, 1968.

64. Lasdon, Leon S., *Optimization Theory for Large Systems*, MacMillian, 1970.

65. Lawden, Derek F., *Analytic Methods of Optimization*, Hafner, 1975.

66. Launberger, D. G., *Optimization by Vector Space Methods*, Wiley, 1969.

67. Leitmann, G., (Editor), *Optimization Techniques with Applications to Aerospace Systems*, Vol. 5., Mathematics in Science and Engineering, Academic press, New York, 1962.

68. Llewellyn, R. W., *Linear Programming*, Holt, Rinehart and Winston, New York, 1964.

69. Lootsma, F. A., (Editor), *Conference on Numerical Methods for Nonlinear Optimization*, Academic Press, New York, 1972.

70. Luenberger, D. G., Introduction to Linear and Nonlinear Programming, Addison-Wesley Publishing Company, Reading, Massachusetts, 1973.

71. Mangasarian, Olvi, L, Nonlinear Programming, McGraw-Hill, 1969.

72. McMillan, Claude, Mathematical Programming, Wiley, 1975.

73. Mitra, G., Theory and Application of Mathematical Programming, Academic Press, 1977.

74. Nemhauser, G. L., Introduction to Dynamic Programming, John Wiley and Sons, Inc., New York, 1967.

75. Neustadt, L. W., Optimization: A Theory of Necessary Conditions, Princeton, U. Press, 1976.

76. Oettli, W. K. and Ritter, K. G., Optimization and Operations Research, Springer-Verlag, 1976.

77. Pierre, Donald A., Optimization Theory with Applications, Wiley, 1969.

78. Plane, D. R. and McMillan, C., Discrete Optimization, Integer Programming and Network Analysis for Management Decisions, Prentice-Hall, Inc. Englewood Cliffs, New Jersey, 1971.

79. Polak, E., Computational Methods in Engineering, A Unified Approach, Vol. 77, Mathematics in Science and Engineering, Academic Press, New York, 1972.

80. Polak, E., Computational Methods in Optimization, Academic Press, 1971.

81. Pontryagin, L. S., et al., The Mathematical Theory of Optimal Processes, Pergmon, 1964.

82. Pun, Lucas, Introduction to Optimization Practice, Wiley, 1969.

83. Reinfeld, N. V. and Vogel, W. R., Mathematical Programming, Prentice-Hall, Inc., Englewood Cliffs, New Jersey, 1958.

84. Rosen, J. B., Mangasarian, O. L., and Ritter, K. (editors), Nonlinear Programming, Academic Press, New York, 1970.

85. Rosen, J. B., The Gradient Projection Method for Nonlinear Programming, Part I, Linear Constraints, Journal of the Society for Industrial and Applied Mathematics, Vol. 8, March 1960, pp. 181-217.

86. Rosen, J. B., The Gradient Projection Method for Nonlinear Programming, Part II, Nonlinear Constraints, Journal of the Society for Industrial and Applied Mathematics, Vol. 9, September 1961, pp. 514-532.

87. Russell, David L., *Optimization Theory*, Benjamin-Cummings, 1970.

88. Rust and Burns, *Mathematical Programming and the Numerical Solution of Linear Equations*, Elsevier, 1972.

89. Sasaki, Kyohei, *Introduction to Finite Mathematics and Linear Programming*, Wadsworth Publ., 1970.

90. Schaefer, Marvin, *A Mathematical Theory of Global Program Optimization*, Prentice-Hall, 1973.

91. Sengupta, J. K., *Stochastic Programming Methods and Applications*, American Elsevier Publishing Company, New York, 1973.

92. Simmonnard, M., *Linear Programming*, translated by Jewell, W. S., Prentice-Hall, Inc. Englewood Cliffs, New Jersey, 1966. (Original French edition: Programmation Lineaire, Dunod, Paris, 1962.)

93. Smith, Donald R., *Variational Methods in Optimization*, Prentice-Hall, 1974.

94. Symthe, W. R., Jr. and Johnson, L. A., *Introduction to Linear Programming With Applications*, Prentice-Hall, Inc., Englewood Cliffs, New Jersey, 1966.

95. Spivey, W. A. and Thrall, R. M., *Linear Optimization*, Holt, Rinehart and Winston, 1970.

96. Stoer, J. and Witzgall, C., *Convexity and Optimization in Finite Dimensions One*, Springer-Verlag, 1970.

97. Szego, C. P. and Szego, G. P., *Minimization Algorithms: Mathematical Theories and Computer Results*, Academic Press, 1973.

98. Tolle, H., *Optimization Methods*, Springer-Verlag, 1975.

99. Vajda, S., *Mathematical Programming*, Addison-Wesley Publishing Company, Reading, Massachusetts, 1961.

100. Vajda, S. *Probabilistic Programming*, Academic Press, New York, 1972.

101. Vajda, S., *Readings in Linear Programming*, John Wiley and Sons, Inc., New York, 1958.

102. Walsh, G. R., *Methods of Optimization*, Wiley, 1975.

103. White, D. J., *Dynamic Programming*, Holden-Day, Inc., San Francisco, 1969.

104. Whittle, Peter, *Optimization Under Constraint Theory and Applications of Nonlinear Programming*, Wiley, 1971.

105. Wilde, Douglass and Beightler, C., *Foundations of Optimization*, Prentice-Hall, 1967.

106. Wilde, D. J., *Optimum Seeking Methods*, Prentice-Hall, Inc., Englewood Cliffs, New Jersey, 1964.

107. Wismer, D. A. (Editor), *Optimization Methods for Large Scale Systems With Applications*, McGraw-Hill Book Company, Inc., New York, 1970.

108. Wismer, D. A., *Introduction to Nonlinear Optimization*, Elsevier, 1977.

109. Wismer, D. A., *Optimization Methods for Large-Scale Systems With Applications*, McGraw, 1971.

110. Zahradmol, Raymond, L., *Theory and Techniques of Optimization for Practicing Engineers*, CBI Pub.

111. Zangwill, Willard, A., *Nonlinear Programming: A Unified Approach*, Prentice-Hall, 1969.

112. Zoutendijk, G., *Mathematical Programming Methods*, Elsevier, 1976.

113. Zoutendijk, G., *Methods of Feasible Directions*, Elsevier Publishing Company, New York, 1960.

114. Zukhovitskii, S. I. and Andreyeva, L. I., *Linear and Convex Programming*, translated by Scripta Technica, Edited by Gelbaum, B. R., W. B. Saunders Company, Philadelphia, 1966.

8.3 PAPERS ON STRUCTURAL OPTIMIZATION

1. Admovich, I. S. and Rikards, R. B., "Discrete Models for Continuous-Type Problems in Design Optimization of Structures," Polymer Mechanics, 12, 5, 752-758 (June 1977). (Translation of Mekhanika Polimerov No. 5, 852-859 (Sept./Oct. 1976) by Consultants Bureau, New York).

2. Admovich, I. S. and Rikards, R. B., "Weight Optimization of an Orthotropic Cylindrical Shell with Variable Properties and a Constraint on Vibration Frequency," Mechanics of Solids, 12, 2, 102-107, (1977), (Translation of Mekhanika Tverdogo Tela, 12, 2, 120-125, (1977), by Allerton Press Inc., New York).

3. Adelman, H. M. and Narayanaswami, R., "Resizing Procedure for Structures Under Combined Mechanical and Thermal Loading," AIAA J. 14, 1484-1486 (1976).

4. Adelman, H. M. and Sawyer, P. L., "Inclusion of Explicit Thermal Requirements in the Optimum Design of Structures," NASA TM X-74017 (1977).

5. Adelman, H. M., Haftka, R. T., and Tsach, U., "Application of Fully Stressed Design Procedures to Redundant and Non-Isotropic Structures," NASA TM-81842, 1980.

6. Adidam, S. R. S., Ramakrisknam, T. S., et al., "A Utility Function Mode for Optimal Material Choice for Multifunctional Performance," Computers and Structures, 8, 5, 583-587 (May 1978).

7. Afimiwala, K. A. and Mayne, R. W., "Evaluation of Optimization Techniques for Applications in Engineering Design," J. Space Craft and Rockets, Vol. 11, No. 10, pp. 673-674, 1974.

8. Agarwal, B. L. and Sobel, L. H., "Weight Comparisons of Optimized Stiffened, Unstiffened, and Sandwich Cylindrical Shells," Journal of Aircraft, 14, 10, 1000-1008, (October 1977).

9. Agaskar, V. L. and Weaver, W. Jr., "Automated Design of Tier Buildings," Computers and Structures, Vol. 12, No. 5-6, pp. 991-1011, 1972.

10. Alblas, J. B., "Optimal Strength of a Compound Column," International Journal of Solids and Structures, 13, 4, 307-320, (1977).

11. Alper, H., Barton, F. W., and McCormick, F. C., "Optimum Design of a Reinforced Plastic Bridge Girder," Computers and Structures, 7, 2, 249-256, (April 1977).

12. Anderson, D. and Salter, J., "Design of Structural Frames to Deflexion Limitations," J. Struct. Eng. Vol. 53, No. 8, pp. 327-333, 1975.

13. Andreev, L., Mossakovskii, V. I., and Obodan, M. I., "On Optimal Thickness of A Cylindrical Shell Loaded by External Pressure," Applied Math. and Mech., Vol. 36, No. 4, pp. 677, 1972.

14. Andries, R. A., Batill, S. M., and Taylor, R. F., "Demonstration and Application of a Minimum-Weight Synthesis Procedure for Flutter Requirements," AFFDL-TM-73-19-FYS.

15. Annamalai, N., Lewis, A. D. M., and Goldberg, J. E., "Cost Optimization of Welded Plate Girders," J. Struct. Div., ASCE, Vol. 98, No. 10, pp, 2235-2246, 1972.

16. Aquilar, R. J., Movassaghi, N., Drewer, J. M., and Forter, J., "Computerized Optimization of Bridge Structures," Computers and Structures, Vol. 3, No. 4, pp. 429-442, 1973.

17. Araslanov, A. M., "Calculation of Uniformly Dependable Minimum Weight Structures," J. Astronautical Sci., Vol. 19, No. 4, pp. 318-325, 1972.

18. Argyris, J. H., Angelopoulos, T., and Bichat, B., "General Method for Shape Finding of Light Weight Tension Structures," Computer Methods Appl. Mech. Eng. Vol. 3, No. 1, pp. 135-149, 1974.

19. Armand, J. L., "Applications of the Theory of Optimal Control of Distributed-parameter Systems to Structural Optimization," CR-2044, NASA (1972).

20. Armand, J. L., "Applications of Optimal Control Theory to Structural Optimization: Analytical and Numerical Approach," Proc. IUTAM Symp. Optimiz. Struct. Des., Springer-Verlag, Warsaw (1975).

21. Armand, J. L. "Minimum-mass Design of a Plate-like Structure for Specified Fundamental Frequency," AIAA J. 9, 1739-1745 (1971).

22. Armand, J. L. and Lodier, B., "Optimal Design of Bending Elements," ASME Energy Technology Conf., Houston, Texas (1977).

23. Arora, J. S., "Survey of Structural Reanalysis Techniques", J. Struct. Div. ASCE, Vol, 102, No. 4, pp. 783-802, 1976.

24. Arora, J. S. and Haug, E. J., Jr., "Efficient Optimal Design of Structures by Generalized Steepest Descent Programming," International Journal for Numerical Methods in Engineering, 10, 4, 747-766, (1976).

25. Arora, J. S., Haug, E. J., Jr., and Rim, K., "Optimal Design of Plane Frames," ASCE J. Struct. Div., Vol. 101, No. 10, pp. 2063-2078, 1975.

26. Arora, J. S. and Govil, A. K., "An Efficient Method for Optimal Structural Design by Substructuring," Computers and Structures, 7, 4, 507-515 (Aug. 1977).

27. Arora, J. S., "Inverse Problem of Structural Optimization," Journal of the Structural Division, Proceedings of the American Society of Civil Engineers, 100, ST, 11, 2355-2360 (Technical Notes), (Nov. 1974).

28. Arora, J. S. and Govil, A. K., "Design Sensitivity Analysis with Substructuring," Journal of Engineering Mechanics Division, ASCE, Vol. 103, No. EM4, Aug. 1977, pp. 537-548.

29. Arora, J. S. and Haug, E. J., "Efficient Hybrid Methods of Optimal Structural Design," Journal of Engineering Mechanics Division, ASCE, Vol. 104, No. EM3, June 1978, pp. 663-680.

30. Arora, J. S. and Haug, E. J., "Methods of Design Sensitivity Analysis to Structural Optimization," AIAA Journal, Vol. 17, No. 9, September 1979, pp. 970-974.

31. Arora, J. S. and Nguyen, D. T., "Eigensolution of Large Structural Systems with Substructures," International Journal for Numerical Methods in Engineering, Vol. 15, No. 3, March 1980, pp. 333-341.

32. Arora, J. S., "Analysis of Optimality Criteria and Gradient Projection Methods for Optimal Structural Design," Computer Methods in Applied Mechanics and Engineering, to appear, 1980.

33. Arora, J. S., Haskell, D. F., and Govil, A. K., "Optimal Design of Large Structures for Damage Tolerance," AIAA Journal, Vol. 18, No. 5, May 1980, pp. 563-570.

34. Arora, J. S., Haug, E. J., and Rajan, S. D., "Efficient Treatment of Constraints in Large-Scale Structural Optimization," Engineering Optimization, to appear 1980.

35. Ashley, H., McIntosh, S. C., Jr., and Weatherill, W. H., "Optimization Under Aeroelastic Constraints," Chap. 11 of Structural Design Applications of Mathematical Programming Techniques, AGARDograph No. 149, Edited by G. G. Pope and L. A. Schmit, 1971.

36. Austin, F., et al., "Aeroelastic Tailoring of Advanced Composite Lifting Surfaces in Preliminary Design," Proc. AIAA/ASME/SAE 17th Structural Dynamics and Materials Conf., Valley Forge, Pennsylvania (1976).

37. Austin, F., "A Rapid Optimization Procedure for Structures Subjected to Multiple Constraints," Paper No. 77-374, AIAA/ASME/SAE 18th Struct. Structural Dynamics and Materials Conf., San Diego, California (1977).

38. Azad, A. K., "Economic Design of Homogeneous I-Beam," Journal of the Structural Division, Proceedings of the American Society of Civil Engineers, 104, ST4, 637-648, (April 1978).

39. Bandyopadhyay, N. and Kappor, M. P., "Optimal Design of Prestressed Concrete Bridge-Girder Section," *Journal of Structural Engineering*, 3, 4, 179-191, (January 1976).

40. Barnett, R. L., "Optimum Prostressed Tubular Columns," *J. Struct. Div.*, ASCE, Vol. 96, No. 2, pp. 291-308, 1976.

41. Bartholomew, P. and Morris, A. J., "Unified Design Approach to Fully Stressed Design," *Eng. Opts.* Vol. 2, No. 1 pp. 3-15 (1978).

42. Batterman, S. C. and Felton, L. P., "Optimal Plastic Design of Doubly Symmetric Closed Structures," *Int. J. Solids Struct.*, Vol. 8, No. 6, pp. 733-750, 1972.

43. Bell, L. C. and Brown, D. M. Guyed, "Tower Optimization," *Computers and Structures*, 6, 6, 447-450, (Dec. 1976).

44. Belytschko, T. B. and Kennedy, J. M., "Computer Models for Subassembly Simulation," *Nuclear Engineering and Design*, 49, 1/2, 17-38, (July 1978).

45. Bennett, J. A., Lin, M. H., and Nelson, M. F., "Application of Optimization Techniques to Problems of Automotive Crash Worthiness," *Int. Cont. on Veh. Struct. Mech.*, 2nd. Southfield, Mich., Apr. 18-20 (1977) SAE Publ. p. 71, pp. 203-210, (1977).

46. Berke L. and Venkayya, V. B., "Review of Optimality Criteria Approaches to Structural Optimization," *Structural Optimization Symposium*, Schmit, L. A., Editor, ASME Winter Annual Meeting 1974, (AMD7).

47. Berke, L., Khot, H. S., Schmit, L. A., Farshi, B., Templeman, A., Sobieszzanski, J., Batt, J. R., and Gellatly, R. A., "Structural Optimization," AGARD Lect. Series No. 70, 1974.

48. Berke, L. and Khot, N. S., "Use of Optimality Criteria Methods for Large Scale Systems," AGARD Lecture Series No. 70, Structural Optimization, 1974.

49. Bhatia, K. G., "An Automated Method for Determining the Flutter Velocity and the Matched Point," *J. Aircraft*, Vol. 11, No. 1, Jan. 1974.

50. Bhatia, K. G., "Rapid Iterative Reanalysis for Automated Design," NASA TN D-7357, Oct. 1973.

51. Biggeis, S. B. and Weigel, Ta. A., "Optimum Cantilever Beams with White Noise Base Excitation," Tall Buildings, Planning, Design and Construction Symp., Proceedings, Vanderbilt Univ., Nashville, Tenn., Nov. 14-15, 1974, pp. 571-584, Published by Vanderbilt Univ., Civil Eng. Prog., 1974.

52. Blachut, J., "Optimum Design of a Flexible Bar by Means of Dynamic Programming," *Mechanika Teorety-Cznai Stosowana*, 15, 1, 125-130, (1977), (in Polish).

53. Block, D. L., "Minimum Weight Design of Axially Compressed Ring and Stringer Stiffened Shells," NASA CR-1766, 1971.

54. Boisserie, J. M. and Glowinski, R., "Optimization of the Thickness Law for Axisymmetric Shells," Computers and Structures, 8, 3/4, 331-343, May 1978).

55. Bond, D., "Optimum Design of Concrete Structures," Eng. Opt., Vol. 1, pp. 17-28, 1974.

56. Borkowski, A., "Optimization of Slab Reinforcement by Linear Programming," Computer Methods in Applied Mechanics and Engineering, 12, 1-17, (Sep. 1977).

57. Boykin, W. H. and Sierakowski, R. L., "Remarks on Pontryagins Maximum Principle Applied to a Structural Optimization Problem," Aeronautical J., Vol. 76, No. 735, pp. 175-176, 1972.

58. Brach, R. M., "Optimum Design of Beams for Sudden Loadings," Proceedings ASCE, Journal of the Engineering Mechanics Div., Vol. 94, No. EM6, Dec. 1968, pp. 1395-1407.

59. Brach, R. M., "On Optimal Design of Vibrating Structures," Journal of Optimization Theory and Applications, Vol. 11, 1973, pp. 662-667.

60. Brant, G. D., "Direct Feasible and Optimal Design of Laterally Unsupported Beams," Engineering Journal, New York, 14, 2, 78-84, (Second Quarter, 1977).

61. Briggs, W. J., "Optimum Design of Frames Using Linear Programming Techniques," thesis presented to Clarkson College of Technology in partial fulfillment of the requirements for the degree of Master of Science, May 1976.

62. Bronowicki, A. J. and Felton, L. P., "Optimum Design of Continuous Thin-Walled Beams," Int. J., Numer. Methods Eng., Vol. 9, No. 3, pp. 711-720, 1975.

63. Brotchie, J. F., Lewis, R. E., and Martin K. G., "Optimization Approach to Viscous Damping of Structures," CSIRO, Div. BLDG. RES., Tech. Paper No. 30, 1972.

64. Brown, R. H., "Minimum Cost Selection of One Way Slab Thickness," J. Struct. Div., Vol. 101, No. 12, pp. 2585-2590, 1975.

65. Brozzetti, J. and Lescourach, Y., "Preliminary Design of Bar Steel Structures Using Statical Method and Linear Pogramming," (in French), RFM, Revue Francaise de Mecanique No. 48, 5014 (1973).

66. Bury, K. V., "Reliability-Constrained Optimum Static Design for Random Load Sequences," Engineering Optimization, 3, 4, 215-220, (September 1978).

67. Cane, V. P. and Harriman, H., "Minimum Cost Design Using a Pre-planned Modal Search Method," Proc. Inst. Civil Engineers, Vol. 59, pt. 2, pp. 237-253, 1975.

68. Capelo, A., "Optimization of a Rigid-Perfectly Plastic Hyperstatic Structure," (in Italian), Bollettino della Unione Matematica Italiana, Serie V 13-B, 1, 55-91 (Apr. 1976).

69. Carmichael, D., "On a Minimum Weight Disk Design Problem," Journal of Applied Mechanics, Transactions of ASME, Series E44, 3, 506-507, (Brief Notes), (Sept. 1977).

70. Carpenter, W. C. and Smith, E. A., "Computational Efficiency in Structural Optimization," J. Eng. Opt. Vol. 1, No. 3, pp. 169-188, 1975.

71. Carpenter, W. C. and Smith, E. A., "A Comment on Powell's Method When Used with SUMT," Engineering Optimization, 2, 4, 279-280 (Technical Note), (1977).

72. Carroll, W. E., "Theorem for Optimum Finite Element Idealizations," Int. J. Solids Struct., Vol. 7, No. 7, pp. 883-895, 1973.

73. Casciati, F., Faravelli, L., and Sacchi, G., "Probabilistic Study of Structures by Linear Programming," (in French), Bulletin Technique de la Suisse Romande, 103, 14, 181-186, (July 1977).

74. Cassis, J. H., "Optimum Design of Structures Subjected to Dynamic Loads," UCLA-ENG-7451, UCLA, School of Engineering and Applied Science, June 1974.

75. Cassis, J. H. and Schmit, L. A., Jr., "Optimum Structural Design with Dynamic Constraints," Journal of the Structural Division, Proceedings of the American Society of Civil Engineers, 102, ST 10, 2053-2071 (Oct. 1976).

76. Cassis, J. H. and Schmit, L. A., Jr., "On Implementation of the Extended Interior Penalty Function," International Journal for Numerical Methods in Engineering IV, 1, 3-23 (1976).

77. Cella, A. and Logcher, R. D., "Automated Optimum Design from Discrete Components," J. Struct. Div., ASCE, Vol. 97, No. 1, pp. 175-190, 1971.

78. Chamis, C. C., "Design of Composite Structural Components," J. Composite Materials, Vol. 8, pp. 231-280, 1975.

79. Chang, D. C. and Barone, M. R., "Structural Optimization in Panel Design," SAE Transactions, 86, 3, 2273, 2281, (1977).

80. Char, C. C., Sun, C. T., and Kohn, S. L., "Strength Optimization for Cylindrical Shells of Laminated Composites," J. Composite Materials, Vol. 9, No. 1, pp. 53-66, 1975.

81. Charrett, D. E. and Rozvany, G. I. N., "Extensions of the Prager-Shield Theory of Optimal Plastic Design," _International Journal of Non-Linear Mechanics_, 7, 1, 51-64, (February 1972).

82. Cheng, F. Y. and Botkin, M. E., "Nonlinear Optimum Design of Dynamic Damped Frames," _Journal of the Structural Division, Proceedings of the American Society of Civil Engineers_, 102, ST 3, 609-627 (Mar. 1976).

83. Cheng, F. Y. and Srifuengfung, D., "Optimum Structural Design for Simultaneous Multicomponent Static and Dynamic Inputs," Paper presented at ASME Energy Technology Conference, Sept. 18-23, 1977, Houston, TX (Ed: V. B. Venkayya).

84. Cheng, W. C. and Mak, C. K., "Computer Analysis of Steel Frame in Fire," _Journal of the Structural Division, Proceedings of the American Society of Civil Engineering_, 101, ST 4, 855-867 (Apr. 1975).

85. Chern, J. M. and Prager, W., "Optimal Design of Trusses for Alternative Loads," _Ing. Arch._, Vol. 41, No. 4, pp. 225-231, 1972.

86. Chern, J. M. and Martin, J. P., "The Multipurpose Optimal Design of Elastic Structures with a Piecewise Uniform Cross Section," _Journal of Applied Mathematics and Physics_, Vol. 22, Fasc. 5, 1971, pp. 834-855.

87. Chern, J. M. and Prager, W., "Optimal Design of Beams for Prescribed Compliance Under Alternative Loads," _J. Opt. Theory Appl._ Vol. 5, No. 6, pp. 424-431, 1970.

88. Chong, K. P., "Optimization of Unstiffened Hybrid Beams," _J. Struct. Div._, ASCE, Vol, 102, No. 2, pp. 401-409, 1970.

89. Chong, K. P. and Harris, D. M., "Automated Design of Continuous Cold Formed Beams," Spec. Conf. on Cold Formed Steel Struct., 2nd, St. Louis, Mo., Rolla, 1973.

90. Chou, T., "Optimal Elastic Design of Rectangular Reinforced Concrete Beam Sections," (in Japanese), _Proceedings of the Japan Society of Civil Engineers No. 250_, 99-109 (June 1976).

91. Chou, T., "Optimum Reinforced Concrete T-Beam Sections," Journal of the Structural Division, _Proceedings of the American Society of Civil Engineers_, 103, ST 8, 1605-1617, (August 1977).

92. Chou, T., "Optimum Design of Reinforced Concrete T-Beam Sections," (in Japanese), _Proceedings of the Japan Society of Civil Engineers No. 258_, 117-132 (Feb. 1977).

93. Chum, Y. W. and Haug, E. J., "Shape Optimization of a Load Carrying Shear Plate," ASME Energy Technology Conf., Houston, Texas (1977).

94. Cohn, M. Z., "Optimal Design of Plastic Structures for Fixed and Shakedown Loading," ASME Paper No. 72 WA/APM-9, 1972.

95. Contro, R., Maier, G., and Zavelani, A., "Inelastic Analysis of Suspension Structures by Nonlinear Programming," *Computer Methods in Applied Mechanics and Engineering*, Vol. 5, No. 2, pp. 127-143, 1975.

96. Craig, R. R., Jr., and Erbug, I. O., "Application of a Gradient-Projection Method to Minimum Weight Design of a Delta Wing with Static and Aeroelastic Constraints," *Computers & Structures*, 6, 6, 529-538 (Dec. 1976).

97. Cwach, E. E. and Stearman, R. O., "Suppression of Flutter on Interfering Lifting Surfaces by the Use of Active Controls," AIAA Paper No. 74-404, AIAA/ASME/SAE 15th Structures, Structural Dynamics, and Materials Conference, Las Vegas, Nevada, April 1974.

98. Datta, T. K., Talib, S. J., and Dixit, V. D., "A Computer Programming Method for Yield Line Analysis," *Journal of Structural Engineering*, 5, 1, 7-15, (April 1978).

99. Davidson, J. W., Felton, L. P., and Hart, G. C., "Optimum Design of Structures with Random Parameters," *Earthquake and Wind Engineering*, UCLA-ENG-7470, EWE 74-04, October 1974.

100. Davidson, J. W., Felton, L. P., and Hart, G. C., "Reliability-Based Optimization for Dynamic Loads," *Journal of the Structural Division*, Proceedings of the American Society of Civil Engineers, 103, ST10, 2021-2035, (Oct. 1977).

101. Dekhtyar, A. S., "Optimization of Rigid Plastic Shell of Revolution," Soviet Applied Mechanics, 13, 5, 473-477, (Nov. 1977), (Translation of *Prikladnaya Mekhanika*, 13, 5, 67-72, (May 1977) by Consultants Bureau, New York).

102. Desmarais, R. N. and Bennett, R. M., "An Automated Procedure for Computing Flutter Eigenvalues," *J. Aircraft*, Vol. 11, No. 2, 1974.

103. Dobbs, M. W. and Nelson, R. B., "Application of Optimality Criteria to Automated Structural Design," *AIAA Journal*, 14, 10, 1436-1443 (Oct. 1976).

104. Douty, R. T. and Crooker, J. O., "Optimization of Long Span Cold Formed Truss Purlins," *J. Struct. Div.*, ASCE, Vol. 100, No. 11, pp. 2275-2288, 1974.

105. Douty, R. T., "Structural Design by Conversational Solution to the Nonlinear Programming Problem," *Computers and Structures*, 6, 4/5, 325-331 (Aug./Oct. 1976).

106. Dwyer, W. J., Emerton, R. K., and Ojalvo, I. U., "An Automated Procedure for the Optimization of Practical Aerospace Structures.

I-Theoretical Development and User's Information, II-Programmer's Manual," AFFDL-TR-70-118-(1971).

107. Dwyer, W. J., "Improved Automated Structural Optimization Program," AFFDL-TR-74-96, 1974.

108. Edwards, L. S., "Optimum Limit State Design of Highway Bridge Superstructures Using Geometric Programming," Engineering Optimization 1, 4, 201-212 (1975).

109. Elmer, C. and Maczynski, J., "On Optimal Shell Prestressing," Journal of Structural Mechanics, 4, 3, 289-305 (1976).

120. Eisenmann, J. R., Kaminske, R. E., Reed, D. L., and Wilkins, D. J., "Toward Reliable Composites: An Examination of Design Methodology," J. Composite Materials, Vol. 7, p. 298, July, 1973.

111. Epstein, H. I. and Alsaigh, J. N., "Minimum Weights for Circular Tube Cantilever Beams," J. Struct. Div., ASCE, Vol. 102, No. 1, pp. 318-322, 1976.

112. Erbatur, F. and Mengi, Y., "On the Optimal Design of Plates for a Given Deflection," Journal of Optimization Theory and Applications, 21, 1, 103-110, (January 1977).

113. Erbatur, F. and Mengi, Y., "Optimal Design of Plates under the Influence of Dead Weight and Surface Loading," Journal of Structural Mechanics, 5, 4, 345-356, (1977).

114. Erbug, I. O., "Application of a Gradient Projection Technique to Minimum-Weight Design of Lifting Surfaces with Aeroelastic and Static Constraints," The Texas Institute for Computational Mechanics, TICOM Report 74-3, June 1974.

115. Farkes, J. and Timar, I., "Optimization of Design by Means of the SUMIT Nonlinear Programming Method," Jarmuvek Mezogazdasagi Gepek, 24, 3, 103-108, (March 1977), (in Hungarian).

116. Farshi, B. and Schmit, L. A., Jr., "Minimum Weight Design of Stress Limited Trusses," Proceedings of the ASCE, Journal of the Structural Division, Vol, 100, ST1, January 1974, pp. 97-107.

117. Faulkner, D., Adamchak, J. C., Snyder, G. J., and Vetter, M. F., "Synthesis of Welded Grillages to Withstand Compression and Normal Loads," Computers and Structures, Vol. 3, No. 2, pp. 221-246, 1973.

118. Felippa, C. A., "Optimization of Finite Element Grids by Direct Energy Search," Applied Math. Mod., Vol. 1, No. 2, pp. 93-96, 1976.

119. Felippa, C. A. and Park, K. C., "Computational Aspects of Time Integration Procedures in Structural Dynamics, Part 1, Implementation," Journal of Applied Mechanics, Transactions of the ASME, 45, 595-602, (September 1978).

120. Felton, L. P. and Nelson, R. B., "Optimized Components in Frame Synthesis," <u>AIAA Journal</u>, Vol. 9, No. 6, pp. 1027-1031, (June 1971).

121. Felton, L. P., "Structural Index Methods in Optimum Design," <u>Structural Optimization Symp., ASME</u> AMD 7 (1974).

122. Feng, T. T., Arora, J. S., and Haug, E. J., Jr., "Optimal Structural Design Under Dynamic Loads," <u>International Journal for Numerical Methods in Engineering</u>, 11, 1, 59-22, (1977).

123. Fleury, C. and Geradin, M., "Optimality Criteria and Mathematical Programming in Structural Weight Optimization," <u>Computers and Structures</u>, 8, 1, 7-17, (February 1978).

124. Fleury C. and Sander, G., "Relationships Between Optimality Criteria and Mathematical Programming in Structural Optimization," Proc. Symp. Applications of Computer Meth. Engng. (Ed. C. Wellford, Jr.), Univ. of Southern California, 507-520 (1977).

125. Foley, M. and Citron, S. J., "A Simple Technique for the Minimum Mass Design of Continuous Structural Members," <u>Journal of Applied Mechanics, Transactions of the ASME</u>, Series E44, 2, 285-290 (June 1977).

126. Fox, R. L., Rao, S. S., and Miura, H., "Automated Design Optimization of Supersonic Airplane Wing Structures Under Dynamic Constraints," Proposed AIAA Paper, Unpublished, (Oct. 1971).

127. Fox, R. L. and Miura, H., "An Approximate Analysis Technique for Design Calculations," <u>AIAA J.</u>, Vol. 9, No. 1, pp. 177-179, (January 1971).

128. Fox, R. L., Miura, H., and Rao, S. S., "Automated Design Optimization of Supersonic Airplane Wing Structures Under Dynamic Constraints," <u>Journal of Aircraft</u>, Vol. 10, No. 6, pp. 321-322, (June 1973).

129. Frangopol, C. and Rondal, J., "Optimum Probability-Based Design of Plastic Structures," <u>Engineering Optimization</u>, 3, 1, 17-25, (June 1977).

130. Frind, E. O. and Wright, P. M., "Gradient Methods in Optimum Structural Design," <u>Journal of the Structural Divisions</u>, Proceedings of the ASCE, Vol. 101, No. ST4, (1975).

131. Fuchs, M. B. and Brull, M. A., "Norm Optimization Method in Structural Design," <u>AIAA Journal</u> 16, 1, 20-28, (January 1978).

132. Fu, K. C., "Application of Search Technique in Truss Configuration Optimization," <u>Computers and Structures</u>, Vol. 3, No. 2, pp. 315-328, (1973).

133. Fulton, R. E., Sobieszczanski, J., and Landrum, E. J., "An Integrated Computer System For Preliminary Design of Advanced Aircraft," AIAA Fourth Aircraft Design, Flight Test, and Operations Meeting, Los Angeles, CA, (Aug. 7-9, 1972).

134. Gajewski, A., "Optimum Design of Physically Nonlinear Cantilever Beam Under Its Own Weight," Rozdrawy Inzynierskie, Engineering Transactions, 24, 3, 453-467, (1976), (in Polish).

135. Gajewski, A., "Minimum Weight Design of a Physically Nonlinear Rotating Rod," (in Polish), Mechanika Teoryczna i Stosowana, 14, 2, 261-271, (1976).

136. Galambos, A. R., Hosain, M. V., and Speirs, W. G., "Optimum Expansion Ratio of Castellated Steel Beams," Engineering Optimization, 1, 4, 213-225, (1975).

137. Gallagher, R. H. and Zienkiewicz, O. C. (Ed.), "Optimization in Structural Design," Wiley, London, (1973).

138. Garg. S., "Derivatives of Eigensolutions for a General Matrix," AIAA Journal, Vol. 11, No. 8, pp. 1191-1194, (August 1973).

139. Gellatly, R. A., Dupree, D. M., and Berke, L., "OPTIM II: a Magic Compatible Large Scale Automated Minimum Weight Design Program," AFFDL-TR-74-97, I and II (1974).

140. Gellatly, R. A. and Berke, L., "Optimal Structural Design," AFFDL-TR-70-165, Air Force Flight Dynamics Laboratory, Wright-Patterson A.F.B., Ohio (1975).

141. Giles, G. L., "Procedure for Automating Aircraft Wing Structural Design," J. Struct. Div., ASCE, Vol 97, No. 1, pp. 99-114, (1971).

142. Giles, Gary L., Blackburn, C. L., and Dixon, S. C., "Automated Procedures for Sizing Aerospace Vehicle Structure (Saves)," J. Aircraft, Vol. 9, No. 12, pp. 812-819, (December 1972).

143. Gjelsvik, A., "Minimum Weight Design of Continuous Beams," Int. J. Solids Struct., Vol. 17, No. 10, pp. 1411-1425, (1971).

144. Goble, G. G. and LaPay, W. S., "Optimum Design of Prestressed Beams," J. Amer. Conc. Inst. Vol. 68, No. 9, pp. 712-718, (1971).

145. Goodall, I. W. and Whitwam, C. M., "On Optimizing Thermal Stresses in Cylindrical Shells," Int. J. Mech. Sci., Vol. 15, No. 1, pp. 99-107, (1973).

146. Gorman, M. R. and Schmidt, L. C., "Heuristic Approach to Automatic Structural Design," Conf. on Comput. in Eng., Proc., Sydney, Aust. (May 16, 17, 1974), pp. 41-45, Publ. by Inst. of Eng. (Natl. Conf. Publ. No. 74/1) Sydney, Aust., (1974).

147. Gorzynski, J. W. and Thornton, W. A., "Variable Energy Ratio Method for Structural Design," ASCE J. Structural Div., Vol. 101, No. 4, pp. 975-990, (1975).

148. Govil, A. K., Arora, J. S., and Haug, E. J., "Optimal Design of Wing Structures with Substructuring," International Journal of Computers and Structures, Vol. 10, pp. 899-910, (December 1979).

149. Govil, A. K., Arora, J. S., and Haug, E. J., "Optimal Design of Frames with Substructuring," International Journal of Computer and Structures, In Press.

150. Griega, I., "Computer Aided Design by means of Finite Elements," European Computer Cont. on Interact. Sys., Proc., London, pp. 155-170, (1975).

151. Grimes, G. C., "Design Techniques and Allowables Criteria for Composite Materials," Sample, Vol. 3, No. 2, pp. 67-76, (January 1972).

152. Grinev, V. B. and Filippov, A. P., "Optimum Circular Plates," Mechanics of Solids, 12, 1, 122-128, (1977), (Translation of Mekhanika Tverdogo Tela, 12, 1, 131-137, (1977), by Allerton Press, Inc., (New York).

153. Grundy, P., "Optimum Layout for Large Areas," J. Struct. Div., ASCE, Vol. 97, No. 7, pp. 2085-2096, (1971).

154. Gunnlaugsson, G. A. and Martin, J. B., "On Optimality Conditions for Trusses with Nonuniform Stress Constraints," J. Struct. Mech. 2, pp. 229-257, (1973).

155. Gura, N. M. and Sierayan, A. P., "Optimum Circular Plate with Constraints on the Rigidity and Frequency of Natural Oscillations," Mechanics of Solids, 12, 1, 129-136 (1977). Translation of Mekhanika Tverdogo Tela 12, 1, 138-145 (1977) by Allerton Press, Inc., New York.

156. Gurvich, I. B., et al., "Weight Optimization of Eccentrically Reinforced Cylindrical Shells," Soviet Applied Mechanics, 13, 7, 724-726, (Jan. 1978), (Translation of Prikladnaya Mekhanica, 13, 7, 113-116, (July 1977), by Consultants Bureau, New York).

157. Gwin, L. B. and Taylor, R. F., "A General Method for Flutter Optimization," AIAA Journal, Vol. 11, No. 12, pp. 1613-1617, (Dec. 1973).

158. Gwin, L. B. and Taylor, R. F., "A General Method for Flutter Optimization," AIAA Paper No. 73-391, AIAA/ASME/SAE 14th Structure, Structural Dynamics, and Materials Conference, Williamsburg, VA, (March 1973).

159. Gwin, L. B., "Optimal Aeroelastic Design of an Oblique Wing Structure," AIAA Paper No. 74-349, AIAA/ASME/SAE 15th Structures, Structural Dynamics, and Materials Conference, Las Vegas, Nevada (April 1974).

160. Gwin, L. B. and McIntosh, S. C., Jr., "A Method of Minimum-Weight Synthesis for Flutter Requirements, Part I - Analytical Investigation," AFFDL-TR-72-22, Part I, (June 1972).

161. Gwin, L. B. and McIntosh, S. C., Jr., "A Method of Minimum-Weight Synthesis for Flutter Requirements, Part II - Analytical Investigation," AFFDL-TR-72-22, Part II, (June 1972).

162. Haeri, H., Lemaire, M., and Cubaud, J. C., "Application of the Finite Element Method and Linear Programming to the Optimization of a Plane Structure," (in French), Annales de l'Institut Technique du Batiment et des Travaux Publics, No. 342, pp. 121-134 (Sept. 1976).

163. Haftka, R. T., "Automated Procedure for Design of Wing Structures to Satisfy Strength and Flutter Requirements," NASA TN D-7262, 7264 (1973).

164. Haftka, R. T. and Yates, E. C., Jr., "On Repetitive Flutter Calculations In Structural Design," AIAA 12th Aerospace Sciences Meeting, Washington, D.C., Paper No. 74-14 (Jan. 1974).

165. Haftka, R. T. and Starnes, J. M., Jr., "WIDOWAC (Wing Design Optimization with Aeroelastic Constraints): Program Manual," NASA TM X-3071, (1974).

166. Haftka, R. T., Starnes, J. M., Jr., and Barton, F. W., "A Comparison of Two Types of Structural Optimization Procedures for Satisfying Flutter Requirements," Presented at the AIAA/ASME/SAE 15th Struct., Structural Dynamics and Materials Conf., American Institute of Aeronautics and Astronautics (1974).

167. Haftka, R. T., "Parametric Constraints with Application to Optimization for Flutter Using a Continuous Flutter Constraint," AIAA J. Vol. 13, pp, 471-475, (1975).

168. Haftka, R. T. and Yates, E. C., Jr., "Repetitive Flutter Calculations in Structural Design," Journal of Aircraft, Vol. 13, pp. 454-461, (1976).

169. Haftka, R. T. and Starnes, J. H., Jr., "Application of a Quadratic Extended Interior Penalty Function for Structural Optimization," AIAA Journal, 14, 6, pp. 718-724, (June 1976).

170. Haftka, R. T., "Optimization of Flexible Wing Structures Subject to Strength and Induced Drag Constraints," AIAA Journal, Vol. 14, pp. 1101-1106, (1977).

171. Haftka, R. T. and Shore, C. P., "Approximate Methods for Combined Thermal/Structural Analysis," NASA TP-1428, (1979).

172. Haftka, R. T., Prasad, G., and Tsach, U., "PARS-Programs for Analysis and Resizing of Structures," NASA CR-159007, (1979).

173. Haftka, R. T. and Prasad, B., "Programs for Analysis and Resizing of Complex Structures," Computers and Structures, Vol. 10, pp. 323-330, (1979).

174. Haftka, R. T. and Prasad, B., "Light Weight Design of the Sides of a Typical 100-Ton High Scale Gondola Car," 3rd International Conference on Vehicle Structural Mechanics, Troy, Michigan, (October 1979).

175. Hall, M. A., "Some Problems of Civil Engineering Analysis and the Theory of (Generalized) Geometric Programming," Eng. Opt. Vol. 2, No. 2, pp. 111-123, (1976).

176. Hansen, H. R., "Application of Optimization Methods," within Structural Optimization Symposium 1974, ASME, Vol. 7, (1974).

177. Hassig, H. J., "An Approximate True Damping Solution of the Flutter Equation by Determinant Iteration," J. Aircraft, Vol. 8, No. 11, (1971).

178. Haug, E. J., Arora, J. S., and Feng, T. T., "Sensitivity Analysis and Optimization of Structures for Dynamic Response," Journal of Mechanical Design Transactions of the ASME, 100, 2, 311-318, (April 1978).

179. Haug, E. J., Jr., Pan, K. C., and Streeter, T. D., "Computational Method for Optimal Structural Design-1, Piecewise Uniform Structures," Int. J. Numer. Methods Engr., Vol. 5, No. 2, pp. 171-184, (1972).

180. Haug, E. J., Pan, K. C., and Streeter, T. D., "A Computational Method for Optimal Structural Design I, Piecewise Uniform Structures," Int. J. Numer. Methods Engr., 10, pp. 747-766 (1976) and 10, pp. 1420-1426 (1976).

181. Haug, E. J. and Arora, J. S., "Design Sensitivity Analysis of Elastic Mechanical Systems," Journal of Computer Methods in Applied Mechanics and Engineering, Vol. 15, pp. 35-62, (1978).

182. Haug, E. J., Jr., Arora, J., S., and Matsul, K., "A Steepest Descent Method for Optimization of Mechanical Systems," Journal of Optimization Theory and Application, Vol. 19, No. 3, pp. 401-424, (1976).

183. Heldenfels, R. R., "Automating the Design Process: Progress, Problems, Prospects, and Potential," AIAA Paper No. 73-410, AIAA/ASME/SAE 14th Structures, Structural Dynamics, and Materials Conference, Williamsburg, VA. (March 1973).

184. Hill, R. D. and Rozvany, G.I.N., "Optimal Beam Layouts the Free Edge Paradox", Journal of Applied Mechanics, Translations of the ASME, Series E44, 4, pp. 696-700, (Dec. 1977).

185. Himmelblau, D. M., "Optimal Design Via Structural Parameters and Nonlinear Programming," Eng. Opt., Vol. 2, No. 1, pp. 17-27, (1976).

186. Hirai, I., Torazo, Y., and Takamura, K., "On a Direct Eigenvalue Analysis for Locally Modified Structures," Int. J., for Numer. Methods In Engineering, Vol. 6, pp. 441-456, Short Communications, (1973).

187. Ho, J. K., "Optimal Design of Multi-Stage Structures, A Nested Decomposition Approach," Computers and Structures, 5, 4, pp. 249-255, (Nov. 1975).

188. Hogley, J. R. and Weatherill, W. H., "SST Technology Follow-On Program-Phase II: A Flutter Analysis Program to Achieve an Optimized Structure for a Supersonic Transport Airplane," FAA-SS-73-13, (Feb. 1974).

189. Holst, O., "Automatic Design of Plane Frames," Denmark Tech. Univ. Research Lab. Dept. No. 53, (1974).

190. Hood, R. V., "Active Controls Changing the Roles of Structural Design," Astronautics and Aeronautics, Vol. 10, No. 8, pp. 50-55, (Aug. 1972).

191. Hornbuckle, J. C. and Boykin, W. H., Jr., "Equivalence of a Constrained Minimum Weight and Maximum Column Buckling Load Problem with Solution," Journal of Applied Mechanics, Transactions of the ASME, 45, 1, pp, 159-164, (Mar. 1978).

192. Hornbuckle, J. C., Nevill, G. E., and Boykin, W. H., "Structural Optimization Using the Finite Element Method Applied to a Beam," Int. J. Numer. Methods Engr., Vol. 19, No. 1 pp. 101-167, (1975).

193. Housner, J. N. and Stein, M., "Flutter Analysis of Swept Wing Subsonic Aircraft with Parameter Studies of Composite Wings," NASA TN D-7539, (1974).

194. Hsiao, M. H., Haug, E. J., Jr., and Arora, J. S., "A State Space Method for Optimal Design of Vibration Isolators," Journal of Mechanical Design, Vol. 101, No. 2, pp. 309-314, (April 1979).

195. Huang, N. C., "Minimum-Weight Design of Vibrating Elastic Structures with Dynamic Deflection Constraint," Journal of Applied Mechanics, Transactions of the ASME, Series E 43, 1 pp. 178-180 (brief notes), (March 1976).

196. Huang, N. C., "Minimum Weight Design of Elastic Cables," Journal of Optimization Theory and Application, 15, 1, pp. 37-49, (Jan. 1975).

197. Hughes, O. F. and Mistree, F., "Some Considerations Regarding Structural Optimization and Finite Element Analysis," Conference on Finite Element Methods in Engineering, University of New South Wales, Sydney, Australia, (Aug. 1974).

198. Isakson, G. and Pardo, H., "ASOP-3: A Program for Minimum-Weight Design of Structures Subjected to Strength and Deflection Constraints," AFFDL-TR-76-157 (1976).

199. Jendo, S., "Optimum Design of Axially-Symmetrical Suspended Structures," Revue Roumaine des Sciences Techniques, Serie de Mecanique Appliquee, 16, 3, pp. 565-582, (1971).

200. Jenkins, W. M., DeJesus, G. C., and Burns, A., "Optimum Design of Welded Plate Girders," Structural Engineer, 55, 12, pp. 547-553, (December 1977).

201. Johnson, E. H., et al., "Optimization of Continuous One-Dimensional Structures Under Steady Harmonic Excitation," AIAA Journal, 14, 12, pp. 1690-1698, (December 1976).

202. Johnson, E. H., "Optimization of Structures Undergoing Harmonic or Stochastic Excitation," Ph.D. Dissertation, Department of Aeronautics and Astronautics, Stanford University (1975) (also NASA CR-142, 936).

203. Jones, R. T., "Application of Multivariable Search Techniques to Structural Design Optimization," NASA CR-2038, (1972).

204. Kamat, M. P. and Simitses, G. J., "Optimal Beam Frequencies by the Finite Element Displacement Method," Int. J. Solids Structs., Vol. 9, No. 3, pp. 415-429, (1973).

205. Kamat, M. P., "Effect of Shear Deformations and Rotary Inertia on Optimum Beam Frequencies," Int. J. Numer. Methods Engr., Vol. 9, No. 1, pp. 51-62, (1975).

206. Kamil, H., "A Computer-Oriented Deterministic-Probalistic Approach for the Extreme Load Design of Complex Structures," Computers and Structures, 6, 4/5, pp. 375-379, (August/October 1976).

207. Kanarachos, A., Koch, M., and Prufer, H. P., "On the Performances Capacity and Economy of Structural Optimization Methods," VDI-Z, 119, 15/16, pp. 759-766, (August 1977), (In German).

208. Karihaloo, B. L. and Niordson, F. I., "Optimum Design of a Circular Shaft in Forward Precision," Proc. IUTAM Symp. Optimiz. Struct. Des., Warsaw, Springer-Verlag (1975).

209. Katarya, R. and Murthy, P. N., "Optimization of Multicell Wings for Strength and Natural Frequency Requirements," J. Computers and Structures, Vol. 5, pp. 225-232, (1975).

210. Kato, B., Nakamura, V., and Anraku, H., "Optimum Earthquake Design of Shear Buildings," ASCE Proceedings, Journal of the Engineering Mechanics Division, Vol. 98, No. EM4, pp. 891-910, (August 1972).

211. Kavlie, D. and Powell, G. H., "Efficient Reanalysis of Modified Structures," *J. Struct. Div., ASCE*, Vol 97, No. 1, pp. 377-392, (1971).

212. Kavlie, D. and Moe, J., "Automated Design of Frame Structures," *J. Struct. Div., American Society Civil Engineers*, Vol. 97, No. ST1, pp. 33-62, (Jan. 1971).

213. Kavanagh, K. T., "Approximate Algorithm for the Reanalysis of Structures by the Finite Element Method," *Computers and Structures*, Vol. 2., No. 5-6, pp. 713-722, (1972).

214. Kelly, D. W., et al., "A Review of Techniques for Automated Structural Design," *Computer Methods in Applied Mechanics and Engineering*, 12, 2, pp. 219-242, (Oct./Nov. 1977).

215. Kelly, D. W., "Dual Formulation for Generating Information About Constrained Optima in Automated Design," *Computer Methods in Applied Mechanical Engineering*, Vol. 15, No. 3, pp. 339-352, (1975).

216. Keshava Rao, M. N., "Optimization of Variable Thickness Plates Under Multiloads," *Journal of Structural Engineering*, 6, 2, pp. 87-92, (July 1978).

217. Khalifa, M. M. K. and Merwin, J. E., "Optimum Plastic Design of Space Frames," *Institution of Civil Engineers, Proceedings, Part 2, Research and Theory*, 63, pp. 769-783, (December 1977).

218. Khan, M. R., Willmert, K. D., and Thornton, W. A., "A New Optimality Criterion Method for Large Scale Structures," AIAA/ASME 19th SDM Conference, Bethesda, Md., pp 47-58, (April 3-5, 1978).

219. Khan, M. R., Thornton, W. A., and Willmert, K. D., "Optimality Criterion Techniques Applied to Mechanical Design," *Journal of Mechanical Design, Transactions of the ASME*, 100, 2, pp. 319-327, (Apr. 1978).

220. Khan, M. R., "Optimization of Trusses," *Journal of Structural Engineering*, 6, 2, pp. 78-86, (July 1978).

221. Khot, N. S., Venkayya, V. B., and Berke, L., "Optimum Design of Composite Structures with Stress and Displacement Constraints," *AIAA Journal*, 14, 2, pp. 131-132, (Feb. 1976).

222. Khot, N. S., et al., "Optimum Design of Composite Wing Structures with Twist Constraint for Aeroelastic Tailoring," *U. S. Air Force Flight Dynamics Laboratory, Technical Report*, AFFDL-TR, pp. 76-117, (December 1976).

223. Khot, N. S., Venkayya, V. B., and Berke, L., "Optimum Structural Design with Stability Constraints," *International Journal for Numerical Methods in Engineering*, 10, 5, pp. 1097-1114, (1976).

224. Khot, N. S., Venkayya, V. B., and et al., "Experience wih Minimum Weight Design of Structures Using Optimality Criteria Methods," SAE Transactions, 86, 3, pp. 2244-2254, (1977).

225. Khot, N. S., "Computer Program (OPTCOMP) for Optimization of Composite Structures for Minimum Weight Design," U.S. Air Force Flight Dynamics Laboratory, Technical Report, AFFDL-TR-76-149, pp. 58, (Feb. 1977).

226. Khot, N. S., Berke, L., and Venkayya, V. B., "Comparisons of Optimality Criteria Algorithms for Minimum Weight Design of Structures," AIAA/ASME 19th SDM Conf., Bethesda, Md., pp. 37-46, (1978).

227. Khot, N. S., Venkayya, V. B., and Berke, L., "Optimum Design of Composite Structures with Stress and Deflection Constraints," AIAA Paper No. 75-141, Represented at AIAA 13th Aerospace Science Meeting, Pasadena, Calif., (1975).

228. Kirsch, U., Reiss, M., and Shamir, R., "Optimum Design by Partitioning into Substructures," J. Struct. Div., American Society of Civil Engineers, Vol. 98, No. 1, pp. 249-268, (1972).

229. Kirsch, U., "Optimum Design of Prestressed Beams," International Journal Computers and Structures, Vol. 2, Pergamon Press, pp. 573-583, (1972).

230. Kirsch U. and Rubinstein, M. F., "Reanalysis for Limited Design Modifications," J. of ASCE, Engineering Mechanics Div., EM2, pp. 61-70, (February 1972).

231. Kirsch, U. and Rubinstein, M. F., "Structural Reanalysis by Iteration," Computers and Structures, Vol. 2, No. 4, pp. 497-510, (1972).

232. Kirsch, U., "Synthesis of Elastic Structures with Controlled Forces," Computers & Structures, 6, 2, pp. 111-116, (April 1976).

233. Kirsch, U., "Optimum Design of Prestressed Plates," J. Struct. Div., ASCE, Vol. 99, No. 7, pp. 1681-1685, (1973).

234. Kirsch, U., "Multilevel Approach to Optimum Structural Design," J. Struct. Div., ASCE, 101, No. ST4, Proc, Paper 11259, pp. 957-974, (1975).

235. Kirsch, U. and Moses, F., "Formulation of Optimal Design in the Behavior Variables Space," Journal of Structural Mechanics, 4, 4, pp, 437-452, (1976).

236. Kirsch, U. and Moses, F., "Optimization of Structures with Control Forces and Displacements," Engineering Optimization, 3, 1, pp. 37-44, (June 1977).

237. Kirsch, U. and Moses, F., "The Relationship Between Plastic and Prestressed Elastic Optimal Design," Mechanics in Engineering,

Proceedings of the First ASCE - EMD Specialty Conference on Mechanics in Engineering held at the University of Waterloo, May 26-28, 1976, Edited by Dubey, R. N. and Lind, N. C., University of Waterloo Press, pp. 207-222, (1977).

238. Kirsch, U. and Moses, F., "Decomposition in Optimum Structural Design," J. of ASCE, Structural Div., Vol. 105, No. ST1, (Jan. 1979).

239. Kirsch, U., "Optimal Design of Trusses by Approximate Compatibility," (to be published), Instr. Journ. Computers and Structures, (1980).

240. Kirsch, U. and Benardout, D., "Optimal Design of Elastic Trusses by Approximate Equilibrium," (to be published), Computer Methods in Applied Mechanics and Engineering, (1980).

241. Kiselev, V. E., "Optimal Design of Nonlinearly Elastic Trusses Under Several Loadings," Stroitelnaya Makhanika i Raschet Sooruzhenii, No. 6, 7-10, (in Russian), (Dec. 1975).

242. Kiusalaas, J. and Reddy, G. B., "DESAP 2-a Structural Design Program with Stress and Buckling Constraints," NASA CR-2797 to 2799 (3 volumes), National Aeronautics and Space Administration, Washington, D. C., (1977).

243. Kowal, Z., "Probabilistic Model Similitude in Limit States of Plates," Bulletin de L'Academie Polanaise Aes Sciences, Serie Des Sciences Techniques, 24, 10, pp. 805-812, (1979).

244. Kowalik, J. S., "Feasible Direction Methods," Chap. 7 of Structural Design Applications of Mathematical Programming Techniques, AGARDograph No. 149, Edited by G. G. Pope and L. A. Schmit, (1971).

245. Krzys, W., "Optimum Design of Thin Walled Closed Cross-Section Columns," Bull. Acad. Pol. Sci., Ser. Sci. Tech., Vol. 121, No. 9, pp. 409-419, (1973).

246. Kufner, V., "Calculation of Limit Load-Carrying Capacity of Frames by Linear Programming," (in Czech), Stavebnicky Casopis 23, 12, pp. 909-923, (Dec. 1975).

247. Kumar, P., "Optimal Force Transmission in Reinforced Concrete Deep Beams," Computers and Structures, 8, 2, pp. 223-229, (April 1978).

248. Kunoo, K. and Yang, T. Y., "Minimum Weight Design of Cyclindrical Shell with Multiple Stiffener Sizes," AIAA Journal, 16, 1, pp. 35-40, (Jan. 1978).

249. Kuzmanovic, B. O. and Willems, N., "Optimum Plastic Design of Steel Frames," J. Struct. Div., ASCE, Vol. 98, No. 8, pp. 1697-1724, (1972).

250. Lai, Y. S. and Achenbach, J. D., "Optimal Design of Layered Structures under Dynamic Loading," Computers and Structures, Vol. 3, No. 3, pp. 559-572, (1973).

251. Lai, Y. S. and Achenbach, J. D., "Direct Search Optimization Method," J. Struct. Div., ASCE, Vol. 99, No. 1, pp. 19-32, (1973).

252. Lamblin, D., Cinquini, C., and Gueriement, G., "Application of Linear Programming to the Optimal Plastic Design of Circular Plates Subject to Technological Constraints," Computer Methods in Applied Mechanics and Engineering, 13, 2, pp. 233-243, (Feb. 1978).

253. Lane, V. P., "Application of Optimization Techniques in Practice," Eng. Opt., Vol. 1, No. 1, pp. 5-15, (1974).

254. Lansing, W., Dwyer, W., Emerton, R., and Ranalli, E., "Application of Fully Stressed Design Procedures to Wing and Empennage Structures," J. of Aircraft, Vol. 8, No. 9, (Sept. 1971).

255. Lansing, W., Lerner E., and Taylor, R. F., "Applications of Structural Optimization for Strength and Aeroelastic Design Requirements," Paper Presented at the 45th AGARD Struct. and Materials Panel Meeting, Voss, Norway, AGARD Report No. 664, (1978).

256. LaPay, W. S. and Goble, G. G., "Optimum Design of Trusses for Ultimate Loads," J. Struct. Div., ASCE, Vol. 97, No. 1, pp. 157-174, (1971).

257. Lasdon, L. S., Fox, R. L., and Ratner, N. W., "An Efficient One-Dimensional Search Procedure for Barrier Functions," Rep. No 46, Div. Solid Mech., Struct. & Mech. Design, Case Western Reserve Univ., (Sept. 1971).

258. Lee, B. S. and Knapton, J., "Optimum Cost Design of a Steel Framed Building," J. Eng. Optimization, Vol. 1, No. 3, pp. 139-153, (1975).

259. Lepik, U., "Minimum Weight Design of Circular Plates with Limited Thickness," Int. J. of Non-Linear Mech., Vol. 17, No. 4, pp. 353-360, (1972).

260. Lepik, U. and Mroz, Z., "Optimal Design of Plastic Structures Under Impulsive and Dynamic Pressure Loading," International Journal of Solid and Structures, 13, 7, pp. 657-674, (1977).

261. Lepik, U., "Optimal Design of Beams with Minimum Compliance," International Journal of Non-Linear Mechanics, 13, 1, pp. 33-42, (1978).

262. Lerner, E. and Markowitz, J., "An Efficient Structural Resizing Procedure for Meeting Static Aeroelastic Design Objectives," AIAA/ASME/19th SDM Conf., Bethesda, Md., pp. 59-66, (1978).

263. Lesdouarch, Y., "Computer Program for the Elastic-Plastic Analysis of Plane Structures," (in French), Construction Metallique, 13, 1, pp. 40-50, (Mar. 1976).

264. Lev, O. E., "A Structural Optimization Solution to a Branch-and-Bound Problem," Quarterly of Applied Mathematics, 34, 4, pp. 365-371, (Jan. 1977).

265. Lev, O. E., "Optimum Choice of Determinate Trusses Under Multiple Loads," Journal of the Structural Division, Proceedings of the American Society of Civil Engineers, 103, ST2, pp. 391-403, (February 1977).

266. Lev, O. E., "On the Solution of a Nonlinear Programming Problem by Decomposition," Journal of Optimization Theory and Applications, 22, 1 pp. 31-34, (May 1977).

267. Lev, O. E., "Optimal Truss Geometry for Two Loading Conditions," Journal of the Structural Division, ASCE, Vol. 104, No. ST8, (August 1978). Presented at the ASCE Second Annual Engineering Mechanics Division Specialty Conference, North Carolina State University, Raleigh, North Carolina, (May 1977).

268. Lev, O. E., "On the Topology and Optimality of Certain Trusses," Journal of the Structural Division, ASCE, in press, (Feb. 1981).

269. Levy, R. and Melosh, R., "Computer Design of Antenna Reflections," AIAA/ASME/SAE 14th Struct., Structural Dynamics and Materials Conf., Williamsburg, V.A., (1973).

270. Levy, H. J. and Wolf, B. M., "Fully Stressed Dynamically Loaded Structures," ASME Publication 74-WA/DE-19, ASME Winter Annual Meeting, (1974).

271. Levy, R., "Computer-Aided Design of Antenna Structures and Components," Computers and Structures, 6, 4/5, pp. 419-428, (Aug./Oct., 1976).

272. Lipson, S. L and Gwinn, L. B., "Discrete Sizing of Trusses for Optimal Geometry," Journal of the Structural Division, Proceedings of the American Society of Civil Engineers, 103, ST5, pp. 1031-1046, (May 1977).

273. Lipson, S. L. and Gwinn, L. B., "The Complex Method Applied to Optimal Truss Configuration," Computers and Structures, 7, 3, pp. 461-468, (June 1977).

274. Lipson, S. L. and Russell, A. D., "Cost Optimization of Structural Roof System," J. Struct. Div., ASCE, Vol. 97, No. 8, pp. 2057-2071, (1971).

275. Lipson, S. L. and Agarwal, U. M., "Weight Optimization of Plane Trusses,", J. Struct, Div., ASCE, 100, No. ST5, Proc. Paper 10521, pp. 865-879, (1974).

276. Liu, S. C. and Neghabat, F., "Cost Optimization Model for Seismic Design of Structures," Bell Syst. Tech. J., Vol. 51, No. 10, pp. 2209-2225, (1972).

277. Lukasiewicz, S. A. and Borajkiewicz, W., "Optimum Design of Elements Introducing a Load into Spherical Shell," Int. J., Solids Struct., Vol. 10, No. 8, (1974).

278. Lure, K. A. and Cherkaev, A. V., "Prager Theorem Application to Optimal Design of Thin Plates," Mechanics of Solids, 11, 6, pp. 139-141, (1976). (Translation of Mekhanika Tverdogo Tela, 11, 6, pp. 157-159, (1976) by Allerton Press, Inc., New York).

279. Lynch, R. W. and Rogers, W. A., "Aeroelastic Tailoring of Composite Materials to Improve Performance," Proc. AIAA/ASME/SAE 17th Struct., Structural Dynamics and Materials Conf. Valley Forge, Pennsylvania, (1976).

280. Maier, G., Zavenlani-Rossi, A., and Benedetti, D., "Finite Element Approach to Optimal Design of Plastic Structures in Plane Stress," Int. J. Numer. Methods Engr. Vol. 4, No. 4, pp. 455-473, (1972).

281. Maier, G., Srinivasar, R., and Save, M. A., "On Limit Design of Frames Using Linear Programming," Journal of Structural Mechanics, 4, 4, pp. 349-378, (1976).

282. Majid, K. I. and Anderson, D., "Optimum Design of Hyperstatic Structures," Int. J. Numer. Methods Engr., Vol. 4, No. 4, pp. 561-578, (1972).

283. Majid, K. I., "Topological Design of Pin Jointed Structures by Non-Linear Programming," Proc. Inst. Civ. Eng. (London), Vol. 55, Part 2, pp. 129-149, (1973).

284. Majid, K. I., "Forces and Deflexions in Changing Structures," Structural Engineer, Vol. 51, No. 3, pp.93-101, (1973).

285. Makowski, M. and Seefer, G., "Optimum Shape Design of a Beam Resting on Elastic Foundation with Normal Stress Restrictions," Mechanika Teoretycznai Stosowana, 15, 3, pp. 359-368, (1977), (in Polish).

286. Marks, W., "Optimization of Beam Cross-Sections," (in Polish), Achiwum Inzynierii Ladowej, 22, 2, pp. 251-263, (1976).

287. Markus, S. S., Oravsky, V., and Simkova, O., "Philosophy of Optimum Design of Damped Sandwich Beams," Acta. Tech. CSAV, Vol. 19, No. 6, pp. 647-662, (1974).

288. Massonnet, C. and Rondal, J., "Optimal Design of Structures: Possibilities and Limitations," (in French), Annales des Travaux Publics de Belgique No. 6, pp. 447-454, (Dec. 1976).

289. Masur, E. F., "Optimal Placement of Available Sections in Structural Eigenvalue Problems," Journal of Optimization Theory and Applications, 15, 1, pp. 69-84, (Jan. 1975).

290. Masur, E. F., "Optimal Structural Design for a Discrete Set of Available Structural Members," Computer Methods in Appl. Mech. and Engr., Vol. 3, No. 2, pp. 195-207, (1974).

291. Mau, S. T. and Sexsmith, R. G., "Minimum Expected Cost Optimization," J. Struct. Div., ASCE, pp. 2043-2058, (1972).

292. McCullers, L. A. and Lynch, R. W., "Dynamic Characteristics of Advanced Filamentary Composite Structures: Vol. II, Aeroelastic Synthesis Procedure Development," AFFDL-TR-73-111.

293. McCullers, L. A. and Lynch, R. W., "Composite Wing Design for Aeroelastic Requirements," Proceedings of the Conference on Fibrous Composites in Flight Vehicle Design, AFFDL-TR-72-130, U. S. Air Force, pp. 951-972, (Sept. 1972). (Available from DDC as AD 907 042L).

294. McCullers, L. A., "Automated Design of Advanced Composite Structures," AMD 7-Struct. Opt. Symp., ASME, New York, pp. 119-133, (1974).

295. McIntosh, S. C., "Structural Optimization via Optimal Control Techniques; A Review," ASME Struct. Optimization Symp., AMD 7, (1974).

296. McKeown, J. J., "Quasi-Linear Programming Algorithm for Optimizing Fiber-Reinforced Structures of Fixed Stiffness," Computer Methods in Applied Mechanical Engineering, Vol. 6, No. 2, pp. 123-154, (1975).

297. McKeown, J. J., "Optimal Composite Structures by Deflection-Variable Programming," Computer Methods in Applied Mechanics and Engineering. 12, 2, 155, 179, (Oct./Nov. 1977).

298. Melchers, R. E., "Optimal Fiber-Reinforced Plate Corners," J. Struct. Div., ASCE, Vol. 99, No. 7, pp. 1681-1685, (1973).

299. Menkes, E. G., Sandor, G. N., and Wang, L. R., "Optimum Design of Composite Multilayer Shell Structures," ASME, Paper No. 71-WA/DE-12, (1971).

300. Miele, A., Mangiavacchi, A., Mohanty, B. P., and Wu, A. K., "Numerical Determination of Minimum Mass Structures with Specified Natural Frequencies," ASME Energy Technology Conf., Houston, Texas, (1977).

301. Mikhailishchev, V. Ya., "Search for Lighest Rod-Type Carrier System by the Dynamic Programming Method," <u>Strengthening of Materials</u>, 3, 11, pp. 98-100, (November 1971), (Translation of <u>Probley Prochnosti</u>, 3, 11, pp. 98-100, (November 1971), by Consultants Bureau, New York.)

302. Miller, R. E., Jr., et al., "Recent Advances in Computerized Aerospace Structural Analysis and Design," <u>Computers and Structures</u>, 7, 1, pp. 315-326, (April 1977).

303. Misteth, E., "Systematic Investigation of the Optimum Safety of Multipurpose Constructions," <u>Muszaki Tudo Many</u>, 52, 1-12, pp. 135-157, (1976), (in Hungarian).

304. Mistree, F. P., "An Automated Design Procedure for the Design of Complex Structures," Ph.D. Thesis, University of California, Berkeley, (1973).

305. Miura, H., "An Optimal Configuration Design of Lifting Surface Type Structures Under Dynamc Constraints,", Rep. No. 48, Div. Solid Mech., <u>Struct. and Mech. Design</u>, Case Western Reserve Univ., (October 1971).

306. Mlejnek, H. P., Bulmeier, J., and Mai, M. M., "Untersuchung and Weiterentwicklung Einiger Verfahren Zur Optimalen Dimensionierung Von Tragwerken," Instut fur Statik und Dynamik Der Luft-und-Raumfahrt-Konstructionen, University of Stuttgart, Report No. BUFT-RVII-TU-3/72B, (1974).

307. Moe, J., "Fundamentals of Optimization," <u>Computers and Structures</u>, Vol. 4, No. 1, pp. 95-113, (1974).

308. Mohraz, B. and Wright, R. N., "Solving Topologically Modified Structures," <u>Computers and Structures</u>, Vol. 3, No. 2, pp. 341-353, (1973).

309. Mole, R. H., "Minimum Weight Structural Configuration of Pin Jointed Truss Cantilevers of Given External Shape," <u>Int. J. Mech. Sci.</u>, Vol. 15, No. 1, pp. 49-63, (1973).

310. Morley, C. T. and Gulvanessian, H., "Optimum Reinforcement of Concrete Slab Elements," <u>Institution of Civil Engineers, Proceedings, Part 2, Research and Theory</u>, 63, pp. 441-454, (June 1977).

311. Morris, A. J., "On Condensed Geometric Programming in Structural Optimization," <u>Computer Methods in Applied Mechanics and Engineering</u>, 15, 2, pp. 139-148, (August 1978).

312. Morris, A. J., "The Optimization of Statically Indeterminate Structures by Means of Approximate Geometric Programming," AGARD Conf. Proc. No. 123, (1973).

313. Morris, A. J., "Structural Optimization by Geometric Programming," *International Journal of Solids and Structures*, 8, 7, pp. 847-864, (July 1972).

314. Morris, A. J., "A Transformation for Geometric Programming Applied to the Minimum Weight Design of Statically Determinate Structures," *International Journal of Mechanical Sciences*, 17, 6, pp. 395-396, (June 1975).

315. Moses, F. and Goble, G., "Automated Optimum Design of Unstiffened Girder Cross Sections," *AISC J.*, Vol. 8, No. 2, pp. 43-47, (1971).

316. Moses, F., "Mathematical Programming Methods for Structural Optimization," ASME Struct. *Optimization Symp.*, AMD 7, (1974).

317. Moses, F. and Goble, G. C., "Practical Applications of Structural Optimization," *J. Struct. Div., ASCE*, Vol. 101, No. 4, pp. 635-648, (1975).

318. Moses, F. and Goble, G. C., "Minimum Cost Structures by Dynamic Programming," *AISC Eng. Journal*, Vol. 7, No. 3, pp. 97-100, (1976).

319. Moses, F., "Structural System Reliability and Optimization," *Computers and Structures*, 7, 2, pp. 283-290, (April 1977).

320. Moses, F. and Kirsch, U., "Decomposition in Optimum Structural Design," *ASCE, J. Struct. Div.*, Vol. 105, No. 1, pp. 85-100, (1979).

321. Motro, R., "Optimization of Space Structures with an Application to Double Layer Grid Frameworks," (in French), *Construction Metallique*, 18, 2, pp. 24-36, (June 1979).

322. Mroz, Z. and Rozvany, G. I. N., "Optimal Design of Structures with Variable Support Conditions," *Journal of Optimization Theory and Applications*, 20, 3, pp. 359-38-, (November 1976).

323. Mroz, Z. and Garstecki, A., "Optimal Design of Structures with Unspecified Loading Distribution," *Journal of Optimization Theory and Applications*, 20, 3, pp. 359-380, (November 1976).

324. Murthy, K. N. and Christiano, P., "Optimal Design for Prescribed Buckling Loads," *J. Struct. Div., ASCE*, Vol. 100, No. 11, pp. 2175-2190, (1974).

325. Mustafa, I., "DESAPI: A Structural Synthesis with Stress and Local Instability Constraints," *Computers and Structures*, 8, pp. 243-256, (1978).

326. Na, T. Y. and Kurajian, G. M., "On Optimal Arch Design," *Journal of Engineering for Industry, Transactions of the ASME Series B.* 99, 1, pp. 37-40, (February 1977).

327. Naman, A. E., "Minimum Cost versus Minimum Weight of Prestressed Slabs," *Journal of the Structural Division, Proceedings of the American Society of Civil Engineers*, 102, ST7, pp. 1493-1505, (July 1976).

328. Nagtegaal, J. C., "Optimal Design of Prestressed Elastic Structures," *Int. J. Mech. Sci.*, Vol. 14, No. 11, pp. 779-781, (1972).

329. Nagtegaal, J. C. and Prager, W., "Optimal Layout of a Truss for Alternative Loads," *Int. J. Mech. Sci.*, Vol. 15, No. 7, pp. 583-592, (1973).

330. Nagtegaal, J. C., "A New Approach to Optimal Design of Elastic Structures," Division of Engineering, Brown University, Providence, R.I., (1973).

331. Nakamura, V., Kato, B., and Anraku, H., "Optimum Earthquake Design of Shear Buildings," *J. Engineering Mech. Div., ASCE Proc.* 98, No. EM4, pp. 891-910, (1972).

332. Nakamura, T. and Nagase, T., "Minimum Weight Design of Multistory, Multispan Plane Frames Subject to Reaction Constraints," *Journal of Structural Mechanics*, 4, 3, pp. 257-287, (1976).

333. Narayanan, S. and Nigan, N. C., "Optimum Structural Design in Random Vibration Environments," *Engineering Optimization*, 3, 2, pp. 97-108, (January 1978).

334. Narusberg, V. L., Rickards, R. B., and Teters, G. A., "Optimization of Reinforced Cylindrical Shells with Nonuniform Thickness," *Polymer Mechanics*, 12, 2, pp. 257-303, (February 1977). (Translation of *Mekhanika Polimerov* No. 2, pp. 298-303, (March/April 1976) by Consultants Bureau, New York.)

335. Narusberg, V. L., "Optimization of a Cylindrical Shell of Reinforced Plastic with a Filler under Dynamic Loading," *Polymer Mechanics*, 13, 5, pp. 737-743, (May 1978). (Translated by Consultants Bureau, New York.)

336. Ng., S. F. and Kulkarni, G. G., "Optimum Design of Longitudinally Stiffened Simply Supported Orthortropic Bridge Decks," *Journal of Sound and Vibration*, 40, 2, pp. 273-284, (May 1975).

337. Nigam, N. C. and Narayanan, S., "Structural Optimization in Aseismic Design," 5th World Conf. Earthq. Engineering, Rome, (1973), (Preprint Paper 374).

338. Niordson, F. I. and Pedersen, P., "A Review of Optimal Structural Design," Proc. 13th Intl. Cong. Theor. Appl. Mech., Moscow, Springer-Verlag, (1973).

339. Noor, A. K. and Lowder, H. E., "Approximate Techniques of Structural Reanalysis," Computers and Structures, Vol. 4, No. 4, pp. 801-812, (July 1974).

340. Noor, A. K., "Multiple Configuration Analysis via a Mixed Method," Journal of the Structural Division, ASCE, (September 1974).

341. Noor, A. K. and Lowder, H. E., "Structural Reanalysis Via A Mixed Method," Computers and Structures, Vol. 5, No. 1, pp. 9-12, (1975).

342. Noor, A. K. and Lowder, H. E., "Approximate Reanalysis Techniques with Substructuring," ASCE J. Str. Div., Vol. 101, No. 8, pp. 1687-1698, (August 1975).

343. Norabhcompipat, T. and Reinschmidt, K. F., "Electronic Computation--Aid to Structural Analysis and Design of Tall Structures," Conf. on Tall Buildings, Proc., Bangkok, Thailand, pp. 414-430, (January 23-25, 1974), sponsored by Asian Inst. of Tech., Bangkok.

344. O'Connell, R. F., "Incremented Flutter Analysis," J. Aircraft, Vol. 11, No. 4, (April 1974).

345. Oda, J., "On a Technique to Obtain an Optimum Strength Shape by the Finite Element Method," Bulletin of the JSME, 20, 140, pp. 160-167, (February 1977).

346. Oda, J. and Yamazaki, K., "On a Technique to an Optimum Strength Shape of an Axisymmetric Body by the Finite Element Method," Bulletin of the JSME, 20, 150, pp. 1524-1532, (December 1977).

347. Ohkubo, S. and Okumura, T., "Graphical Optimization of Steel Girders Based on Suboptimization of Girder Elements," Proceedings of the Japan Society of Civil Engineers, No. 252, pp. 23-34, (in Japanese), (August 1976).

348. Olhoff, N., "Optimal Design of Vibrating Rectangular Plates," Int. J. Solids and Struct., 10, pp. 93-109, (1974).

349. Olhoff, N., "Survey of the Optimal Design of Vibrating Structural Elements: Part I," Theory Shock and Vibration Digest, 8, 8, pp. 3-10, (August 1976).

350. Olhoff, N., "Survey of the Optimal Design of Vibrating Structural Elements: Part II," Applications Shock and Vibration Digest, 8, 9, pp 3-10, (September 1976).

351. O'Neil, H. M., "Computerized Design of Post-Tensioned Continuous Beams and Flat Plates," J. Prestressed Concr. Inst., Vol. 18, No. 3, pp. 42-50, (1973).

352. Paczkowski, W., "Optimization of a Simply Supported Reinforced Concrete Beam," Archiwum Inzynierii Ladowej, 22, 4, pp. 559-569, (in Polish), (1976).

353. Pandit, G. S. and Sharma, S. C., "Optimum Steel Distribution in Corner Slabs," Journal of the Structural Division, Proceedings of the American Society of Civil Engineers, 102, ST12, pp. 2360-2365, (Technical Notes), (December 1976).

354. Pape, G. and Thierauf, G., "Computer-Aided Dimensioning of Structures: Design and Over All Stability Computer-Oriented Design of Structures," Konstruktiever Ingenieurbau Berichte, No. 28, pp. 5-27, (in German), (April 1977).

355. Pappas, M. and Amba-Rao, C. L., "A Direct Search Algorithm for Automated Optimum Structural Design," AIAA J. 9, pp. 387-393, (1971).

356. Pappas, M., "Use of Direct Search in Automated Optimal Design," Journal of Engineering for Industry, Transactions of the ASME Series B, 94, 2, pp. 395-401, (May 1972).

357. Pappas, M. and Allentuch, A., "Structural Synthesis of Frame Reinforced Submersible Circular Cylindrical Shells," Computerss and Structures, Vol. 4, No. 2, pp. 253-280, (1974).

358. Pappas, M. and Allentuch, A., "Extended Capability for Automated Designs of Frame Stiffened Submersible Cylindrical Shells," Computers and Structures, Vol. 4, No. 5, pp. 1025-1029, (1974).

359. Park, K. C. and Felippa, C. A., "Computational Aspects of Time Integration Procedures in Structural Dynamics, Part 2, Error Propagation," Journal of Applied Mechanics, Transaction of the ASME, 45, 3, pp. 603-611, (September 1978).

360. Patnaik, S. and Dayaratnam, P., "Behaviour and Design of Pin Connected Structures," Int. J. Numerical for Methods in Engineering, Vol. 2, No. 4, pp. 579-595, (1971).

361. Patnaik, S. and Srivastava, N. K., "On Automated Optimum Design of Trusses," Journal of Structural Engineering, 3, 4, pp. 164-178, (January 1976).

362. Patnaik, S. and Sankaran, G. V., "Optimum Design of Stiffened Cylindrical Panels with Constraint on the Frequency in the Presence of Initial Stress," International Journal for Numerical Methods in Engineering, 10, 2, pp. 283-299, (1979).

363. Pedersen, P., "On the Optimal Layout of Multi-Purpose Trusses," Computers and Structures, Vol. 12, No. 5-6, pp. 695-712, (1972).

364. Pedersen, P., "Optimal Joint Positions for Space Trusses," J. Structural Division, ASCE, Vol. 99, No. 12, pp. 2459-2476, (1973).

365. Pectu, V., "Compatibility of the Plastic Hinge Rotations in the Optimal Plastic Design of Reinforced Concrete Continuous Beams," <u>Studii Si Cercetari De Mecanica Aplicata</u>, 30, 3, pp. 645-672, (in Romanian), (1971).

366. Petrova, I. S. and Rikards, R. B., "Optimization of a Member with Variable Modulus of Elasticity," <u>Polymer Mechanics</u>, 10, 2, pp. 237-242, (October 1975). (Translation of Mrkhanika Polimerov, 10, 2, pp. 277-284 by Consultants Bureau, New York, (March/April 1974)).

367. Petukhov, L. V. and Troitskii, V. A., "Some Optimal Problems of the Theory of Longitudinal Vibrations of Rods," <u>Appl. Math. Mech.</u>, Vol. 36, No. 5, pp. 842-851, (1972).

368. Phansalkar, S. R., "Matrix Iterative Methods for Structural Reanalysis," <u>Computers and Structures</u>, Vol. 4, No. 4, pp. 779-880, (1974).

369. Phoa, Y. T., "Vibcal - A Code for Computation of Swept Wing Bending and Torsional Vibration Modes and Frequencies," Aerophysics Research Corp. Technical Note Tn-105, (August 1971).

370. Phoa, Y. T., "An Automated Flutter Solution Procedure - Afsp," Aerophysics Research Corporation Technical Note Tn-109, (1971).

371. Phoa, Y. T., "A Computerized Flutter Solution Procedure," National Symposium on Computerized Structural Analysis and Design, Washington D.C., (march 1972).

372. Phoa, Y. T., et al., "Application of a Rand-Developed Nonlinear Programming Method to Flutter Optimization," Nat. Symp. on Computerized Structural Analysis and Design, Washington D.c., (March 29-31, 1976).

373. Pickett, R. M., Jr., Rubenstein, M. F., and Nelson, R. B., "Automated Structural Synthesis Using a Reduced Number of Design Coordinates," <u>AIAA Journal</u>, Vol. 11, No. 4, pp. 489-494, (April 1973).

374. Pierson, B. L., "A Survey of Optimal Structural Design Under Dynamic Constraints," <u>Int. J. Numerical for Methods in Engineering</u>, 4, pp. 491-499, (1972).

375. Pierson, B. L., "Further Discrete Variable Results for a Panel Flutter Optimization Problem," <u>Int. J. Numerical for Methods in Engineering</u>, Vol. 7, No. 4, pp. 537-543, (1973).

376. Pines, S. and Newman, M., "Structural Optimization for Aeroelastic Requirements," Proc. AIAA/ASME/SAE 14th Struct., Structural Dynamics and Material Conf., Williamsburg, VA, (1973).

377. Pines, S. and Newman, M., "Constrained Structural Optimization for Aeroelastic Requirements," <u>J. Aircraft</u>, Vol. 11, No. 6, (June 1974).

378. Pister, K. S., "Mathematical Modeling for Structural Analysis and Design," <u>Nuclear Eng. and Des.</u>, Vol. 18, No. 3, pp. 353-375, (1972).

379. Pluntett, R., "Optimum Damping Distribution for Structural Vibration," <u>U.S. Nav. Res. Lab., Shock Vil. Bull.</u>, No. 42, Part 4, pp. 57-64, (1972).

380. Pochtman, Yu, M. and Khariton, L. E., "Optimal Design for a Structure Considering Reliability," <u>Stroitel Naya Mekhanika i Raschet Sooruzhenii</u>, No. 6, pp. 8-15, (in Russian), (December 1976).

381. Polizzotto, C., "Optimum Plastic Design for Multiple Sets of Loads," <u>Meccanica</u>, Vol. 9, No. 3, pp. 203-213, (1974).

382. Polovinkin, A. I., "Automatic Search for Engineering Structures of Optimal Design," <u>Eng. Dybern.</u>, Vol. 9, No. 5, pp. 809-817, (1971).

383. Pope, G. G. and Schmit, L. A. (Ed.), "Structural Design Applications of Mathematical Programming Methods," <u>AGARD Second Symp. Struct. Opt.</u>, AGARDograph No. 149, (1971).

384. Popelar, C. H., "Optimal Design of Beams Against Buckling: A Potential Energy Approach," <u>Journal of Structural Mechanics</u>, 4, 2, pp. 181-196, (1976).

385. Popelar, C. H., "Optimal Design of Structures Against Buckling: A Complementary Energy Approach," <u>Journal of Structural Mechanics</u>, 5, 1, pp. 45-66, (1977).

386. Prager, W., "Optimal Layout of Cantilever Trusses," <u>Journal of Optimization Theory and Applications</u>, 23, 1, pp. 111-117, (September 1977).

387. Prager, W., "Optimal Design of Statically Determinate Beams for Given Deflection," <u>Int. J. Mech. Sci.</u>, Vol. 13, No. 10, pp. 893-895, (1971).

388. Prager, W., "Optimality Criteria in Structural Design," AGARD Report No. 538, (December 1971).

389. Prager, W., "Conditions for Structural Optimality," <u>Computers and Structures</u>, Vol. 2, pp. 833-840, (1972).

390. Prager, W., "Minimum-Weight Design of a Statically Determinate Truss Subject to Constraints on Compliance, Stress, and Cross-Sectional Area," <u>J. Appl. Mech., Trans ASME</u>, Vol. 40, Series E, No. 1, pp. 313-314, (1973).

391. Prasad, B. and Haftka, R. T., "A Cubic Extended Interior Penalty Function for Structural Optimization," <u>Int. Journal for Numerical Methods in Engineering</u>, Vol. 14, pp. 1107-1126, (1979).

392. Prasad, B. and Haftka, R. T., "Structural Optimization with Plate Finite Elements," Journal of the Structural Division, ASCE, Vol. 105, No. ST11, pp. 2367-2382, (1979).

393. Rabbat, B. G. and Collins, M. P., "The Computer Aided Design of Structural Concrete Sections Subjected to Combined Loading," Computers and Structures, 7, 2, pp. 229-236, (April 1977).

394. Ragsdell, K. M. and Phillips, D. T., "Optimal Design of a Class of Welded Structures Using Geometric Programming," ASME, Paper 75, DET-86, (1975).

395. Raj, P. P. and Durrant, S. O., "Optimum Structural Design by Dynamic Programming," Journal of the Structural Division, Proceedings of the American Society of Civil Engineers, 102, ST8, pp. 1575-1589, (August 1979).

396. Ramanathan, R. K., "A Multilevel Approach in Optimum Design of Structures Including Buckling Constraints," Ph.D. Thesis, University of California, Los Angeles, (1976).

397. Ramesh, C. K. and Karve, S. R., "Optimization for Stiffened Plates -Some Studies," Journal of Structural Engineering, 3, 4, pp. 201-211, (January 1976).

398. Rand, R. A. and Shen, C. N., "Optimum Design of Composite Shells Subject to Natural Frequency Constraints," J. Computers and Structures, Vol. 3, No. 2, pp. 247-263, (1973).

399. Rao, S. S., "Rates of Change of Flutter Mach Number and Flutter Frequency," AIAA Journal, Vol. 10, No. 11, pp. 1526-1528, (November 1972).

400. Rao, S. S., "Optimization of Complex Structures to Satisfy Static, Dynamic and Aeroelastic Requirements," International Journal for Numerical Methods in Engineering, 8, 2, pp. 249-269, (1974).

401. Rao, S. S., "Optimum Design of Structures Under Shock and Vibration Environment," Shock and Vibration Digest, 7, 12, pp. 61-70, (December 1975).

402. Rao, S. S., "Structural Optimization Under Comb. Blast and Acoustic Loading," AIAA Journal, 14, 2, pp. 276-278, (Technical Notes), (February 1976).

403. Rao, G. V., Shore, C. P., and Narayanaswami, R., "An Optimality Criterion for Resizing Heated Structures with Temperature Constraints," NASA TN D-8525, (1977).

404. Redd, L. T., Gilman, J., Jr., Cooley, D. E., and Sevart, F. D., "A Wind-Tunnel Investigation of a B-52 Model Flutter Suppression System," AIAA Paper No. 74-401, AIAA/ASME/SAE 15th Structures, Structural Dynamics and Materials Conference, Las Vegas, Nev., (April 1974).

405. Reinschmidt, K. F., "Discrete Structural Optimization," J. Struct. Div., ASCE, 97, No. ST1, pp. 133-156, (1971), (Proc. Fifth Conf. Electronic Computation).

406. Reinschmidt, K. F. and Russell, A. D., "Applications of Linear Programming in Structural Layout and Optimization," J. Computers and Struct., 4, pp. 855-869, (1974).

407. Reinschmidt, K. F. and Norabhoompipat, T., "Structural Optimization by Equilibrium Linear Programming," ASCE J. Struct. Div., Vol. 101, No. 4, pp. 855-869, (1974).

408. Reiss, R., "Minimal Weight Design for Conical Shells," J. Appl. Mech., Trans. ASME, Series E., Vol. 41, No. 3, pp. 599-603, (1974).

409. Rikards, R. B., "Optimization Models Including Statistical Parameters for Shells Made of Composite Materials," Polymer Mechanics, 12, 6, pp. 916-924, (July 1977), (Translation of Mekhanika Polimerov, No. 6, pp. 1048-1058, (November/December 1976) by Consultants Bureau, New York).

410. Rizzi, D., "Optimization of Multi-constrained Structures Based on Optimality Criteria," Proc. AIAA/ASME/SAE 17th Struct., Structural Dynamics and Materials Conf., King of Prussia, Pennsylvania, pp. 448-462, (1976).

411. Rodrigues, J. S. N., "Node Numbering Optimization in Structural Analysis," ASCE J. Struct. Div., Vol. 101, No. 3, pp. 361-376, (1975).

412. Rohde, S. M. and McAllister, G. T., "Some Representations of Variations with Application to Optimization and Sensitivity Analysis," International Journal of Engineering Sciences, 16, 7, pp. 443-449, (1978).

413. Rohrle, H., "Structural Dynamic Optimization of Satellites," Proc. Third Symp. Large Struct. for Manned Spacecraft, Frascati, Italy, (1973).

414. Roger, K. L., Felt, L. R., and Hodges, G. E., "Active Flutter Suppression - A Flight Test Demonstration," AIAA Paper No. 74-402, AIAA/ASME/SAE 15th Structures, Structural Dynamics and Materials Conference, Las Vegas, Nev., (April 1974).

415. Romstad, K. M., Hutchingson, J. R., and Runge, K. H., "Design Parameter Variation and Structural Response," Int. Journal of Numerical Methods for Engineering, Vol. 5, pp. 337-349, (1973).

416. Rosenblueth, E., "Optimum Design for Infrequent Disturbances," Journal of the Structural Division, Proceedings of the American Society of Civil Engineers, 102, ST9, pp. 1807-1825, (September 1976).

417. Rosenblueth, E. and Asfura, A., "Optimum Seismic Design of Linear Shear Buildings," J. Struct. Div., ASCE, Vol. 102, No. 5, pp. 1077-1084, (1976).

418. Rosenblueth, E., "Towards Optimum Design Through Building Codes," J. Struct. Div., ASCE, Vol. 102, No. 3, pp. 591-608, (1976).

419. Rowan, Jack C. and Burns, Thomas A., "Aeroelastic Loads Predictions Using Finite Element Aerodynamics," AIAA 12th Aerospace Sciences Meeting, Washington, D.C., AIAA Paper 74-106, (January 30-February 1, 1974).

420. Rozvany, G. I. N., "Slabs with Variable Straight Reinforcement," J. Struct. Div., ASCE, Vol. 97, No. 5, pp. 1521-1532, (1971).

421. Rozvany, G. I. N. and Adidam, S. R., "Absolute Minimum Volume of Reinforcement in Slabs," Journal of the Structural Division, Proceedings of American Society of Civil Engineers, 98, ST5, pp. 1217-121, (Technical Notes), (May 1972).

422. Rozvany, G. I. N., "Basic Geometrical Properties of Optimal Flexural Force Transmission Fields," Journal of Structural Mechanics, 2, 4, pp. 259-264, (1973).

423. Rozvany, G. I. N. and Ganagahanaiah, C., "Grillages of Least Weight-Simply Supported Boundaries," Int. J. Mech. Sci., Vol. 15, No. 8, pp. 665-677, (1973).

424. Rozvany, G. I. N., "Optimization of Unspecified Generalized Forces in Structural Design," J. Applied Mechanics, ASME, Vol. 41, Series E, No. 4, pp. 1143-1145, (1974).

425. Rozvany, G. I. N., "Analytical Treatment of Some Extended Problems in Structural Optimization," Parts I and II, J. Struct. Mech. 3, pp. 359-402, (1974-1975).

426. Rozvany, G. I. N. and Mroz, Z., "Optimal Design Taking Cost of Joints into Account," Journal of the Engineering Mechanics Division, Proceedings of the American Society of Civil Engineers, 101, EM6, pp. 917-921, (Technical Notes), (December 1975).

427. Rozvany, G. I. N. and Hill, R. D., "The Theory of Optimal Load Transmission by Flexure," Advances in Applied Mechanics, 16, pp. 183-308, (1976).

428. Rozvany, G. I. N. and Prager, W., "Optimal Design of Partially Discretized Grillages," Journal of the Mechanics and Physics of Solids, 24, 2/3, pp. 125-136, (June 1976).

429. Rozvany, G. I. N., "Optimal Design of Multiload Multispan Systems," Journal of the Engineering Mechanics Division, Proceedings of the American Society of Civil Engineers, 102, EM6, pp. 1085-1087, (Technical Notes), (December 1976).

430. Rozvany, G. I. N. and Mroz, Z., "Column Design: Optimization of Support Conditions and Segmentation," *Journal of Structural Mechanics*, 5, 3, pp. 279-290, (1977).

431. Rozvany, G. I. N., "Optimal Plastic Design: Allowance for Self-Weight," *Journal of the Engineering Mechanics Division, Proceedings of the American Society of Civil Engineers*, 103, EMG, pp. 1165-1170, (Technical Notes), (December 1977).

432. Rudisill, C. S. and Bhatia, K. G., "Optimization of Complex Structures to Satisfy Flutter Requirements," *AIAA J.* 9, pp. 1487-1491, (1971).

433. Rudisill, C. S. and Bhatia, K. G., "Second Derivatives of the Flutter Velocity and the Optimization of Aircraft Structures," *AIAA Journal*, Vol. 10, No. 12, pp. 1569-1572, (December 1972).

434. Rudisill, C. S. and Cooper, J. L., "An Automated Procedure for Determining the Flutter Velocity," *J. Aircraft*, Vol. 10, No. 7, pp. 442-444, (1973).

435. Rudisill, C. S., "Derivatives of Eigenvalues and Eigenvectors for a General Matrix," *AIAA Journal*, Vol. 12, No. 5, pp. 721-722, (May 1974).

436. Runge, K. H., "Evaluation of Automated Economical Design Methods for Multistory Steel Frameworks," Ph.D. thesis, University of California, Davis, (1972).

437. Safin, R. K., "Optimal Design of Arched Roofs," (in Russian), *Issiedovaniya Po Teorii Plastin i Obolochek No. 10*, pp. 407-411, (1973).

438. San, P. F., Arora, J. S., and Haug, E. J., Jr., "Fail Safe Optimal Design of Framed Structures Subjected to Earthquakes," *Eng. Opt.*, Vol. 2, No. 1, pp. 65-71, (1976).

439. Sanford, M. C., Abel, I., and Gray, D. L., "A Transonic Study of Active Flutter Suppression Based on an Aerodynamic Energy Concept," AIAA Paper No. 74-403, AIAA/ASME/SAE 15th Structures, Structural Dynamics and Materials Conference, Las Vegas, Nev., (April 1974).

440. Sander, G. and Fleury, C., "A Mixed Method in Structural Optimization," ASME Energy Technology Conf., Houston, Texas, (1977), (see Reference 38).

441. Save, M. A., "General Criterion for Optimal Structural Design," *J. Opt. Theory Appl.*, Vol. 15, No. 1, pp. 119-129, (January 1975).

442. Sawczuk, A. and Mroz, Z., Editors, *Optimization in Structural Design*, Symposium Warsaw/Poland, August 21-24, 1973; Berlin, Springer-Verlag, (1975).

443. Schilling, C. G., "Optimum Properties of I-Shaped Beams," *Journal of the Structural Division, Proceedings of the American Society of Civil Engineerss*, 100, ST12, pp. 2385-2401, (December 1974).

444. Schmit, L. A., Jr., "Structural Synthesis 1959-1969: A Decade of Progress," in Recent Advances in Matrix Methods of Structural Analysis and Design, University of Alabama Press, Huntsville, (1971).

445. Schmit, L. A., Jr., "Literature Review and Assessment of the Present Position," Chapt. 4 of *Structural Design Applications of Mathematical Programming Techniques*, AGARDograph No. 149, Edited by G. G. Pope and L. A. Schmit, Jr., (1971).

446. Schmit, L. A., Jr. and Farshi, B., "Some Approximation Concepts for Structural Synthesis," AIAA Paper No. 73-341, AIAA/ASME/ASE 14th Structures, Structural dynamics and Materials Conference, Williamsburg, VA, (March 1973).

447. Schmit, L. A., Jr., (Editor), *Structural Optimization Symposium*, American Society of Mechanical Engineers, Vol. 7, (1974).

448. Schmit, L. A., Jr. and Farshi, B., "Some Approximation Concepts for Structural Synthesis," *AIAA Journal*, Vol. 12, No. 5, pp. 692-699, (May 1974).

449. Schmit, L. A., Jr. and Miura, H., "A New Structural Analysis/Synthesis Capability-Access 1," *AIAA Journal*, 15, 9, pp. 661-671, (May 1976).

450. Schmit, L. A., Jr. and Tash Kandi, M. A., "Influence of Temperature Change on Optimal Laminate Design," *AIAA Journal*, 15, 9, pp. 1238-1241, (September 1977).

451. Schmit, L. A., Jr. and Farshi, B., "Optimum Design of Laminated Fiber Composite Plates," *International Journal for Numerical Methods in Engineering*, 11, 4, pp. 623-640, (1977).

452. Schmit, L. A., Jr., "Structural Synthesis from Abstract Concept to Practical Tool-Synopsis," Proc. AIAA/ASME 18th Struct., Structural Dynamics and Materials Conf., San Diego, California, (1977).

453. Schmit, L. A., Jr. and Ramanathan, R. K., "Multilevel Approach to Minimum Weight Design Including Buckling Constraints," *AIAA Journal*, 16, 2, pp. 97-104, (February 1978).

454. Schuldt, S. B., et al., "Application of a New Penalty Function Method to Design Optimization," *Journal of Engineering for Industry, Transactions of the ASME*, Series B99, 1, pp. 31-36, (February 1977).

455. Schulz, D., "Optimization and Design of the Rear Fuselage of the A 300 B Aircraft Structures," *AGARD 2nd Symposium on Structural Optimization*, Milan, Italy, (April 1973).

456. Seaburg, P. A. and Salmon, C. G., "Minimum Weight Design of Light Gage Members," *J. Struct. Div., ASCE*, Vol. 97, No. 1, pp. 203-222, (1971).

457. Segenreich, S. A. and McIntosh, S. C., "Weight Minimization of Structures for Fixed Flutter Speed via as Optimality Criterion," Proc. AIAA/ASME/SAE 16th Struct., Structural Dynamics and Materials Conf., Denver, Colorado, (1975).

458. Segenreich, S. A., Johnson, E. H., and Rizzi, P., "Three Contributions to Minimum Weight Structural Optimization with Dynamic and Aeroelastic Constraints," Stanford University, SUDAAR Report No. 501, (1976).

459. Segenreich, S. A., Zouain, N. A., and Herskovits, J., "An Optimality Criteria Method Based on Slack Variables Concept for Large Scale Structural Optimization," *Proc. Symp. Applications of Computer Methods in Engineering*, (Ed. C. Wellford, Jr.), University of Southern California, pp. 563-572, (1977).

460. Selleri, F. and Spadallini, O., "Optimal Design of Prestressed Plane Cable Structures," *Journal of Structural Mechanics*, 5, 2, pp. 179-205, (1977).

461. Shamie, J. and Schmit, L. A., Jr., "Frame Optimization Including Frequency Constraints," *ASCE J. Struct. Div.*, Vol. 101, No. 1, pp. 283-293, (1975).

462. Sheu, C. Y. and Schmit, L. A., Jr., "Minimum Weight Design of Elastic Redundant Trusses Under Multiple Static Loading Conditions," *AIAA Journal*, 10, 2, pp. 155-162, (February 1972).

463. Shirk, M. H. and Griffin, K. E., "The Role of Aeroelasticity in Aircraft Design with Advanced Filamentary Composite Materials," paper presented at the Second Conference on Fibrous Composites in Flight Vehicle Design, Dayton, Ohio, (May 1974).

464. Siegel, S., "A Flutter Optimization Program for Aircraft Structural Design," Proc. AIAA 4th Aircraft Design, Flight Test and Operations Meeting, Los Angeles, California, (1972).

465. Simitses, G. J. and Ungbhakorn, V., "Weight Optimization of Stiffened Cylinders Under Axial Compression," *Computers and Structures*, 5, 5/6, pp. 305-314, (December 1975).

466. Simitses, G. J. and Kotras, T., "The Optimal Euler-Bernoulii Cantilever," *Journal of the Engineering Mechanics Division, Proceedings of the American Society of Civil Engineers*, 101, EM6, pp. 922-929, (Technical Notes), (December 1975).

467. Simitses, G. J. and Giri, J., "Minimum Weight Design of Stiffened Cylinders Subjected to Pure Torsion," *Computers and Structures*, 7, 5, pp. 667-677, (October 1977).

468. Simitses, G. J. and Aswani, M., "Minimum Weight Design of Stiffened Cylinders Under Hydrostatic Pressure," *Journal of Ship Research*, 21, 4, pp. 217-224, (December 1977).

469. Simodynes, E. E., "Gradient Optimization of Structural Weight for Specified Flutter Speed," *Journal of Aircraft*, Vol. 11, No. 3, pp. 143-147, (March 1974).

470. Singaraj, N. M. and Sridhar Rao, J. K., "Optimization in Trusses Using Optimal Control Theory," *Journal of the Structural Division, Proceedings of the American Society of Civil Engineers*, 101, ST5, pp. 1037-1051, (May 1975).

471. Smith, E. A., "Minimum Weight Design of Non Uniform Beams," *ASCE J. Struct. Div.*, Vol. 105, No. 7, pp. 1559-1564, (1979).

472. Smith, P. G. and Wilson, E. L., "Automatic Design of Shell Structures," *J. Struct. Div., ASCE*, pp. 191-201, (1971).

473. Sohieszczanski, J. and Loendorf, D., "A Mixed Method for Automated Design of Fuselage Structures," AIAA/ASME/SAE 13th Structures, Structural Dynamics and Materials Conference, San Antonio, Texas, (April 10-14, 1972).

474. Solnes, J. and Holst, O. L., "Optimization of Framed Structures Under Earthquake Loads," 5th World Conf. Earthquake Engineering, Rome, (1973), (preprint paper 376).

475. Spillers, W. R., "Note on the Design of Beams," *J. Appl. Mech. Trans. ASME*, Vol. 38, Sec. B, No. 4, pp. 1073-1074, (1971).

476. Spillers, W. R. and Lev, O. E., "Design for Two Loading Conditions," *Int. J. Solids Structures*, Vol. 7, pp. 1261-1267, (1971).

477. Spillers, W. R. and Friedland, I., "On Adaptive Structural Design," *J. Struct. Div., ASCE*, Vol.. 98, No. 10, pp. 2155-2164, (1972).

478. Spillers, W. R. and Al-Banna, S., "Optimization Using Iterative Design Techniques," *Computers and Structures*, Vol. 3, No. 2, pp. 1263-1271, (1973).

479. Spillers, W. R., "On the Relationship Between Buckling and Optimal Structural Design," *Journal of the Franklin Institute*, 29, 6, pp. 463-466, (Brief Communication), (June 1975).

480. Spillers, W. R., "Iterative Design for Optimal Geometry," *Journal of the Structural Division, ASCE*, Vol. 101, ST7, Proc. Paper 11439, pp. 1435-1442, (July 1975).

481. Stachowicz, A., "Optimum Design of Subgrade Water as a Problem in Multi-level Optimization," *Archinnm Inzynierii Ladowej*, 22, 4, pp. 537-558, (in Polish), (1976).

482. Stadler, W., "Uniform Shallow Arches of Minimum Weight and Minimum Maximum Deflection," Journal of Optimization Theory and Applications, 23, 1, pp. 137-165, (September 1977).

483. Starnes, J. H., Jr. and Haftka, R. T., "Preliminary Design of Composite Wings for Buckling, Stress, and Displacement Constraints," Journal of Aircraft, Vol. 16, pp. 564-570, (1979).

484. Starnes, J. H., Jr. and Haftka, R. T., "Preliminary Design of Composite Wing-Box Structures for Global Damage Tolerance," Paper presented at the AIAA/ASME/ASCE/AHS 21st Structures, Structural Dynamics and Material Conference, Seattle, Washington, (May 1980).

485. Storaasli, O. O. and Sobieszczanski, J. E., "Design Oriented Structural Analysis," AIAA Paper No. 73-338, AIAA/ASME/SAE 14th Structures, Structural Dynamics and Materials Conference, Williamsburg, VA, (March 1973).

486. Stroud, W. J., Dexter, C. B., and Stein, M., "Automated Preliminary Design of Simplified Wing Structures to Satisfy Strength and Flutter Requirements," NASA TN D-6534, (December 1971).

487. Sun, P. F., Arora, J. S., and Haug, E. J., Jr., "Fail-Safe Optimal Design of Structures," Engineering Optimization, 2, 1, pp. 43-53, (1976).

488. Surtees, J. O. and Tordoff, D., "Optimum Design of Composite Box Girder Bridge Structures," Institution of Civil Engineers, Proceedings, Part 2, Research and Theory, 63, pp. 181-194, (March 1977).

489. Switsky, H., "Design of Finite Element Structures with Deflection Constraints," Fairchild Republic Co., Sa140N1314, (May 1975).

490. Tabak, E. I., "Generalized Optimality Criteria for Structural Design," thesis presented to the University of Toronto, in 1977, in partial fulfillment of the requirements for the degree of Doctor of Philosophy.

491. Taig, I. E. and Kerr, R. I., "Optimization of Aircraft Structures with Multiple Stiffness Requirements," Second Symposium on Structural Optimization, AGARD-CP-123, Milan, Italy, (April 1973).

492. Taleb-Agha, G. and Nelson, R. B., "Method for the Optimum Design of Truss-Type Structures," AIAA Journal, 14, 4, pp. 436-445, (April 1976).

493. Taylor, J. E., "Minimum Mass Bar for Axial Vibration at Specified Natural Frequency," AIAA Journal, Vol. 5, pp. 1911-1913, (1967).

494. Taylor, J. E., "On the Production of Structural Layout for Maximum Stiffness," *J. Opt. Theory and Appl.*, Vol. 15, No. 1, pp. 145-155, (January 1975).

495. Taylor, J. E. and Rossow, M. P., "Optimal Truss Design Based on Algorithm Using Optimality Criteria," *International Journal of Solids and Structures*, 13, 10, pp. 913-923, (1977).

496. Taylor, R. F. and Gwin, L. B., "Application of a General Method for Flutter Optimization," *Proceedings of the Second Symposium on Structural Optimization*, AGARD-CP-123, pp. 13-1 - 13-13, (April 1973).

497. Templeman, A. B., "Optimum Truss Design by Sequential Geometric Programming," *Journal of Structural Engineering*, 3, 4, pp. 155-163, (January 1976).

498. Terai, K., "Application of Optimality Criteria in Structural Synthesis," UCLA-ENG-7446, (June 1974).

499. Thakkar, M. C. and Pulsari, B. S., "Optimal Design of Prestressed Concrete Poles," *J. Struct. Div., ASCE*, Vol. 98, No. 1, pp. 61-74, (1972).

500. Thakkar, M. C. and Rao, J. K., "Optimal Design of Prestressed Concrete Pipes Using Linear Programming," *Computers and Structures*, Vol. 4, No. 2, pp. 373-380, (1974).

501. Thermann, K., "Optimal Design Criteria of Dynamically Loaded Elastic Structures," *Proc. IUTAM Symp. Optimiz. Struct. Des.*, Warsaw, Springer-Verlag, (1975).

502. Thomas, H. R., Jr. and Brown, D. M., "Optimum Least Cost Design of a Truss Roof System," *Computers and Structures*, 7, 1, pp. 13-22, (February 1977).

503. Thompson, J. M. T., "Erosion of Optimum Designs by Compound Branching Phenomena," *J. Mech. Phys. Solids*, Vol. 21, No. 3, pp. 135-144, (1973).

504. Thornton, W. A. and Majumdar, D. K., "Reanalysis Information for Eigen-Values Derived from a Differential Equation Analysis Formulation," *AIAA J.*, Vol. 12, No. 12, pp. 1749-1751, (1974).

505. Toakley, A. R. and Williams, D. G., "The Optimum Design of Stiffened Panels Subject to Compression Loading," *Engineering Optimization*, 2, 4, pp. 239-250, (1977).

506. Tocher, J. L. and Karnes, R. N., "The Impact of Automated Structural Optimization on Actual Design," AIAA Paper No. 73-361, (1971).

507. Tokarz, B., "A Direct Method for Optimum Design for Flat Slabs," (in German), *Bautechnik* 53, 11, pp. 374-379, (November 1976).

508. Trikha, D. N. and Murthy, N. S., "Semi-Automatic Optimization Approach for Reinforced Concrete Frame Spacing by Limit Analysis," Journal of Structural Engineering, 6, 2, pp. 96-101, (July 1978).

509. Triplett, W. E. and Ising, K. D., "Computer Aided Stabilator Design Including Aeroelastic Constraints," Journal of Aircraft, Vol. 8, No. 7, pp. 554-561, (July 1971).

510. Troitskii, V. A., "Optimization of Elastic Bars in the Presence of Free Vibrations," Mechanics of Solids, 11, 3, pp. 139-146, (1976), (Translation of Mekhanka Tverdogo Tela, 11, 3, pp. 145-152, (1976) by Allerton Press, Inc., New York).

511. Twisdale, L. A. and Khachaturian, N., "Absolute Minimum Weight Structures by Dynamic Programming," J. Struct. Div., ASCE, Vol. 99, No. 11, pp. 2339-2344, (1973).

512. Twisdale, L. A. and Khachaturian, N., "Multistage Optimization of Structures," J. ASCE Struct. Div., Vol. 101, No. 5, pp. 1005-1020, (1975).

513. Vachajitpan, P. and Rockey, K. C., "Design Method for Optimum Unstiffened Girders," Journal of the Structural Division, Proceedings of the American Society of Civil Engineers, 104, ST1, pp. 141-155, (January 1978).

514. Vanderplaats, G. N. and Moses, F., "Automated Design of Trusses for Optimum Geometry," Proc. ASCE, J. Struct. Div. 98, ST6, pp. 671-690, (1972).

515. Vanderplaats, G. N. and Moses, F., "Structural Optimization by Methods of Feasible Directions," Computers and Structures, Vol. 3, No. 4, pp. 739-755, (1973).

516. Vanderplaats, G. N., "CONMIN A Fortran Program for Constrained Function Minimization: User's Manual," NASA TMX-62, 282, (1973).

517. Vanderplaats, G. N., "Design of Structures with Optimum Geometry," NASA Technical Memorandum, NASA TMX-62462, 12, pp., (August 1975).

518. Vanderplaats, G. N., "Structural Optimization via a Design Space Hierarchy," International Journal for Numerical Methods in Engineering, 10, 3, pp. 713-717, (short communications), (1976).

519. Vanderplaats, G. N., "Structural Analysis and Design Program, Users Manual," NASA-TMX (in preparation), NASA Ames Research Center, Moffett Field, CA, (1976).

520. Vankeuren, G. M., Jr. and Eastep, F. E., "Use of Galerkin's Method for Minimum-Weight Panels with Dynamic Constraints," Journal of Spacecraft and Rockets, 14, 7, pp. 414-418, (July 1977).

521. Venkayya, V. B., "Design of Optimum Structures," J. Computers and Struct., 1, pp. 265-309, (1971).

522. Venkayya, V. B., Khot, N. S., Tischler, V. A., and Taylor, R. F., "Design of Optimum Structures for Dynamic Loads," Third Conf. Matrix Meth. Struct. Mech., Wright-Patterson AFB, Ohio, pp. 619-658, (1971).

523. Venkayya, V. B. and Khot, N. S., "Design of Optimum Structures to Impulse Type Loading," AIAA J. 13, pp. 989-994, (1973).

524. Venkayya, V. B., Khot, N. S., and Berke, L., "Application of Optimality Criteria Approaches to Automated Design of Large Practical Structures," Second. Symp. Struct. Opt. AGARD-CP-123, Ilan, Italy, (1973).

525. Venkayya, V. B. and Khot, N. S., "Structural Optimization," Struct. Mech. Comp. Progs., Symp., Proc. Univ. of MD., June 1975, Publ. by Univ. Press of VA. Charlottesville, (1974).

526. Venkayya, V. B. and Khot, N. S., "Design of Optimum Structures to Impulse Type Loading," AIAA Journal, 13, 8, pp. 989-994, (August 1975).

527. Venkayya, V. B. and Cheng, F. Y., "Resizing of Frames Subjected to Ground Motion," Proc. Int. Symp. on Earthquake Structural Engineering, St. Louis, MO, (August 19-21, 1976). Publ. by Univ. of Mo., Dep. of Civil Eng., Rolla, Vol. 1, pp. 597-612, (1976).

528. Venkayya, V. B. (Ed.), "Structural Optimization Methods--Parts I and II", A collection of papers presented at ASME Energy Technology Conf. and Exhibition, Houston, Texas, (1977).

529. Venkayya, V. B. and Tischler, V. A., "OPTSTAT - A Computer Program for Optimal Design of Structures Subjected to Static Loads," AFFDL-TR-79-(in preparation).

530. Venkateswara, R. G., Shore, C. P., and Narayanaswami, R., "An Optimality Criterion for Sizing Members of Heated Structures with Temperature Constraints," NASA Technical Note, NASA TN D-8525, 44 pp., (October 1977).

531. Vepa, K., "Generalization of an Energetic Optimality Condition for Non-conservative Systems," J. Struct. Mech. 2, pp. 229-257, (1973).

532. Vitiello, I., "Standardization and Optimum Structural Design by Dynamic Programming," Journal of Optimization Theory and Application, 23, 1, 183, 191, (September 1977).

533. Vrouwenveider, A. C. W. M., "Optimum Fire Resistance," Heron, 23, 3, 19, pp., (1977).

534. Waddoups, M. E., Jackson, S. K., and Rogers, C. W., "The Integration of Composite Structures Into Aircraft Design," J. Composite Materials, Vol. 6, pp. 174-190, (April 1972).

535. Wallace, D. and Sireg, A. S., "Optimum Design of Prismatic Bars Subjected to Longitudinal Impact," J. Eng. Industry, Trans. ASME, Vol. 93, Series B, No. 2, pp. 659-666, (1971).

536. Walsh, P. F., "Algorithm for Displacement Sensitivities for Structural Design," Proc. Conf. on Computing in Engineering, Sydney, Aust., pp. 226-228, (May 16-17, 1974).

537. Wang, C. H., "Optimum Design of Prestressed Concrete Members," J. Struct. Div., ASCE, Vol. 96, No. 7, pp. 1525-1534, (1976).

538. Warner, W. H., "Optimal Design in Axial Motion for Several Frequency Constraints," J. Opt. Theory Appl, Vol. 15, No. 1, pp. 157-166, (January 1975).

539. Weisshaar, T. A., "Approximate Solutions to Idealized Structural Dynamic Optimization Problems," Journal of Optimization Theory and Applications, 16, 1/2, pp. 119-133, (July 1975).

540. Wells, A. A., "Composite Materials and the Designer," Composites, (May 1972).

541. Wilby, C. A., "Optimization of Design of Circular Tanks," Institution of Civil Engineers, Proceedings, Part 2, Research and Theory, 63, pp. 921-924, (Technical Notes), (December 1977).

542. Wilkinson, K., Lerner, E., and Taylor, R. F., "Practical Design of Minimum-weight Aircraft Structures for Strength and Flutter Requirements," J. Aircraft, 13, pp. 614-624, (1976).

543. Wilkinson, I., Markowitz, J., Lerner, E., George, D., and Batill, S., "FASTOP: A Flutter and Strength Optimization Program for Lifting-Surface Structures," J. Aircraft, 14, pp. 581-587, (1977).

544. Wright, P. M. and Feng, C. C., "Optimum Design of Frames Using a Multimode Scheme," The Engineering J., Trans. Eng. Institute of Canada 14, pp. 1-6, (1971).

545. Yamada, Y. and Daiguji, H., "Optimum Design of Cable Stayed Bridges Using Optimality Criteria," (in Japanese), Proceedings of the Japan Society of Civil Engineers, No. 253, pp. 1-12, (September 1976).

546. Yamakawa, H. and Okumura, A., "Optimum Design of Structures with Regard to Their Vibrational Characteristic: First Report, A General Method of Optimum Design," Bulletin of the JSME. 19, 138, pp. 1458-1466, (December 1976).

547. Yang, J. C. S. and Tsui, C. Y., "Optimum Design of Structures of Composite Materials in Response to Aerodynamic Noise and Noise

Transmission," NASA Contractor Report, NASA CR-155332, 49 pp., (December 1977).

548. Yang, W. H., "On a Class of Optimization Problems for Framed Structures," Computer Methods in Applied Mechanics and Engineering, 15, 1, pp. 85-97, (July 1978).

549. Yoshida, N., Burkett, S., and Cotter, T. P., "Structural Sensitivity Analysis with NASTRAN," NASA-MSPC-MAF, Chrysler Corporation, Space Div., (1973).

550. Zagajeski, S. W. and Berkero, V., "Optimum Seismic-Resistant Design of R/C Frames," ASCE J. Struct. Div., Vol. 105, No. 5, pp. 829-845, (1979).

551. Zarghamee, M. S., "Minimum Weight Design with Stability Constraint," J. Struct. Div., ASCE, Vol. 96, No. 8, pp. 1697-1710, (1976).

552. Zaripov, N. G., "Optimizing Finned Pilotless Flight Vehicle Design Parameters," Soviet Aeronautics, 20, 4, pp. 90-93, (1977).

553. Zavelani, A., "A Compact Linear Programming Procedure for Optimal Design in Plane Stress," Journal of Structural Mechanics, 2, 4, pp. 301-324, (1973).

8.4 COMPUTER PROGRAMS ON STRUCTURAL OPTIMIZATION

It is almost inconceivable to make any meaningful numerical computation in structural optimization without the aid of electronic computers. It is therefore implicitly understood that computer programs were written and used to obtain the numerical results, which are reported in most of the references on structural optimization, including the ones listed in this report.

By comparison the number of structural optimization computer programs, which are documented and available for universal use in design and analysis, is very small. During the preparation of this report, information on such programs was publicly solicited. Many engineers and researchers responded to this call with programs which did not satisfy these requirements.

The list given in this section contains only programs which are documented (with a User's Manual) and publicly available.

Cheng, F. Y., Srifuengfung, D., and Sheng, L. H., "ODSEWS-2D Optimum Design of Static, Earthquake, and Wind Steel Structures," Technical Report for the National Science Foundation, 1980, Available at the National Technical Information Service, U.S. Department of Commerce, Virginia.

Douty, R., "Design of Steel Connections by Math Programming," Journal of the Structural Division, ASCE, Vol. 106, No. ST5, Proc. Paper 15397, May 1980, p. 1135-1154.

Dwyer, W. J., Emerton, R. K., and Ojalvo, I. U., "An Automated Procedure for the Optimization of Practical Aerospace Structures. I-Theoretical Development and User's Information, II-Programmer's Manual," AFFDL-TR-70-118 (1971).

Dwyer, W. J., "An Immproved Automated Structural Optimization Program," AFFDL-TR-75-96, September 1974.

Gellatly, R. A., Dupree, D. M., and Berke, L., "OPTIM II: A Magic Compatible Large Scale Automated Minimum Weight Design Program" AFFDL-TR-97, Vols. I and II, July 1974.

Haftka, R. T. and Starnes, J. H., Jr., "WIDOWAC (Wing Design Optimization with Aeroelastic Constraints): Program Manual," NASA TM X-3071, 1974.

Haftka, R. T., Prasad, B., and Tsach, U., "PARS-Programs for Analysis and Resizing of Structures--User Manual," NASA CR-159007, April 1979.

Isakson, G., Pardo, H., Lerner, E., and Venkayya, V. B., "ASOP-3: A Program for the Optimum Design of Metallic and Composite Structures Subjected to Strength and Deflection Constraints," AIAA Paper 77-378, presented at AIAA/ASME 18th Structures and Structural Dynamics and Materials Conference, San Diego, March 1977.

Khot, N. S., "Computer Program (OPTCOMP) for Optimization of Composite Structures for Minimum Weight Design," AFFDL-TR-76-149, February 1977.

Lynch, R. W. and McCullers, L. A., "Wing Aero-Elastic Synthesis Procedure," AFFDL-TR-73-111, March 1973.

Mustafa, I., "DESAPI: A Structural Synthesis with Stress and Local Instability Constraints," Computers and Structures, 8, pp. 243-256, (1978).

Schmit, L. A. and Miura, H., "A New Structural Analysis/Synthesis Capability ACCESS 1," AIAA Journal, Vol. 14, No. 5, pp. 661-671, May 1976.

Schmit, L. A. and Miura, H., "An Advanced Structural Analysis/Synthesis Capability ACCESS 2," International Journal for Numerical Methods in Engineering, Vol. 12, 1978, pp. 353-357.

Schmit, L. A., Jr. and Fleury, C., "An Improved Analysis/Synthesis Capability Based on Dual Methods ACCESS 3," AIAA Paper 79-0721, presented at 20th Structures, Structural Dynamics and Materials Conference, St. Louis, MO, April 1979.

Somekh, E. and Kirsch, U., "Structural Design Using Interactive Optimization," Proceedings, Seventh Conference on Electronic Computation, St. Louis, August 1979, pp. 168-189.

Vanderplaats, G. N., "CONMIN - A Fortran Program for Constrained Function Minimization: User's Manual," NASA TM X-62, 282, 1973.

Vanderplaats, G. N., "SADT" (Structural Analysis and Design of Trusses), a computer program developed at NASA Ames Research Center, March 1979.

Venkayya, V. B. and Tischler, V. A., "OPSTAT - A Computer Program for Optimal Design of Structures Subjected to Static Loads," AFFDL-TR-80.

Wilkinson, K., Markowitz, J., Lerner, D. George, and Batill, S. M., "FASTOP: A Flutter and Strength Optimization Program for Lifting Surface Structures," J. Aircraft, 14, pp. 581-587, (1977).

PART III
ANALYTICAL METHODS

9. DISTRIBUTED PARAMETER STRUCTURES

9.1 INTRODUCTION

A multitude of papers have appeared since 1970 in the fields of optimal control, function space optimization, nonlinear programming, finite dimensional structural optimization, and to a lesser degree continuous material distribution in optimization of structural elements. The relationship among these fields in finite dimensional design problems (selection of a parameter vector in R) is becoming reasonably well understood and developed. In the areas of infinite dimensional or distributed parameter optimization of structures (selection of a function of one or more design variables describing continuous distribution of material over a 1, 2, or 3 dimensional structure), the interrelationship of control theory, function space optimization, and structural optimization is far less clear. It is the purpose of this review to analyze the literature that has appeared since 1970 in the field of distributed parameter structural design, particularly as regards the method employed and applications of these methods to specific classes of structural optimization problems.

A substantial literature on optimal design of elastic structure involving continuous distribution of a design variable over one or two space dimensions has occurred. Applications have generally been to test problems involving bars, shafts, beams, and plates of variable section. A variety of performance constraints have been placed on the design problems, which serve to segregate the classes of problems. Research in structural optimization is largely uncoupled from a massive effort in mathematics and science on optimization of distributed parameter systems.

A brief review of books published on structural optimization, review articles that have appeared, and applicable distributed parameter optimization theory is given in this section. The detailed review of distributed parameter structural optimiation literature is then organized according to problem type. This classification is not unique, but is consistent with categorization of objective and constraint that has been followed by authors of preceding reviews. Papers are cited in the reference list in the order they are encountered in the problem classification used in this paper. They are referenced in later sections as they relate to another problem class.

It is intended that this review should be comprehensive for the period 1970 to 1980. However, exhaustiveness cannot be claimed because of the massive literature, particularly Soviet literature that has not yet been translated. Selected papers that appear prior to 1970 are cited to provide correlation with previous work.

---

*This chapter is based on a lecture by E. G. Haug, presented at NATO Advanced Study Institute in Iowa City, in May-June 1980. This work has been supported by NSF Project ENG 77-19967.

This review includes only a few papers on methods based on nonlinear programming solution of discretized model of structures. Some literature is cited in which a finite dimensional variable space is employed with a differential equation model of structural response. To compensate for this deficiency and to cover other aspects of analytical structural optimization, such as plastic design, Appendix A has been included. This appendix refers the reader to a reprint of a review by G. I. N. Rozvany and Z. Mroz, which was published as a feature article in Applied Mechanics Review in 1977. An update by G. I. N. Rozvany is also included in this appendix. The present chapter in combination with Appendix A may be considered as a comprehensive review of analytical methods. Finally, the rather special literature on nonconservatively loaded structures has not been included. Even with this restriction in scope, substantial literature has appeared in the 1970's.

9.1.1 Literature Reviews

The first comprehensive literature review on structural optimization appeared in 1963 by Wasiutynski and Brandt (1). They present a comprehensive review of literature through approximately 1962, citing 234 references. Sheu and Prager (2) continue with a comprehensive review through approximately 1967, citing an additional 146 references. These reviews consider nonlinear programming and distributed parameter methods and include plastic structural optimization. They are the last comprehensive reviews that have appeared in the growing field of structural optimization.

In 1972, Pierson (3) presented a review on structural design under dynamic constraints, with emphasis on structural vibration, flutter, and transient dynamic response. In 1973, Niordson and Pedersen (4) gave a critical and mathematical review of structural optimization literature from approximately 1968 to 1972, dealing with both mathematical programming and analytical methods of structural optimization. McIntosh (5) reviewed applications of optimal control techniques applied to structural optimization, through approximately 1974. Primary emphasis in this review is on structures that are modeled with ordinary differential equations and constraints on static, dynamic, and eigenvalue response.

In a sequence of reviews in the Shock and Vibration Digest, Olhoff (6) and Ranacharyulu and Done (7) review literature on dynamic structural optimization. Olhoff's reviews (6) focus primarily on vibration response, citing literature through 1976. The more recent review (7) treat both vibration and transient response constraints, through approximately 1978. Both of these reviews address numerical and analytical aspects of distributed parameter optimal structural design.

For reviews of related structural optimization problems, the reader is referred to review by Venkayya (8), where primary emphasis is on finite dimensional structural optimization through 1977, and to a recent review by Weisshaar and Plaut (9) on optimization of structure under nonconservative loading.

9.1.2 Books

Prager (10) presented a series of lectures on structural optimization in 1974 that provides an insightful introduction to the field of distributed parameter structural optimization. His primary emphasis is on optimality criterion. The most substantial text on distributed parameter optimal design is the proceedings of a symposium held in Warsaw, Poland in 1973 (11), edited by Sawczuk and Mroz. This book contains a wealth of valuable papers, many of which will be discussed later in this review. In 1976, Rozvany (12) published a text on optimal design of flexural systems, with primary emphasis on design for plastic response, but with a chapter on elastic response optimization. Finally, Haug and Arora (13) present a text on numerical methods of structural optimization, with an introduction to distributed parameter optimization.

9.1.3 Applicable Distributed Parameter Optimal Control Literature

The massive literature on distributed parameter optimal control is primarily oriented to classes of problems that cannot be related to the mainstream of structural design optimization. For example, much of the distributed parameter control literature is on linear control systems. The structural optimization problem is nonlinear, however, since the design characteristics to be selected appear in the coefficients of the otherwise linear differential equations. When taken as equations involving both state and design variables, the equations of solid mechanics are necessarily nonlinear. Further, most optimal control literature involves evolution equations (initial-value problems) and seeks a feedback, or time dependent solution of the optimization problem. The structural design optimization problem, however, may be viewed as a control problem in which the control is time independent, or open loop. This situation is of little interest to most researchers in the field of optimal control and is given very little attention. Finally, the massive literature on existence theory, sufficiency theory, identification theory, and much of the theory of necessary conditions is of little assistance in either theoretical or computational solution of structural design problems.

Robinson (14) surveyed the distributed parameter control literature covering over 200 papers during the period 1966 through 1969. In his review of open loop control techniques, which are most closely allied with the type of problem encountered in structures, Robinson observes that if the necessary conditions can be stated in integral equation form, then a number of Soviet generated techniques are available. However, if the equations are in differential equation form, some methods are available to obtain solutions, but not with any degree of generality. Robinson's closing remark is that the techniques developed apply only to a very restricted class of problems and, in the general case, the prospects of determining solutions from necessary conditions are not good.

Of value in application of necessary conditions to problems of distributed parameter optimal design is a paper by Lurie (15), in which elliptic partial differential equations of the kind encountered in structural optimization are reduced to first order form and general necessary conditions are derived for nonlinear optimization. The necessary conditions presented there have served as the theoretical foundation for some of the structural optimization work reviewed here. It may be noted that the even ordered partial differential equations of conservative continuum mechanics can generally be put in self-adjoint form. Taking advantage of this fact, it is possible to obtain necessary conditions by the same techniques employed by Lurie. The resulting necessary conditions involve an adjoint variable that is the solution of the same basic differential operator equation, but with a different nonhomogeneous term.

As noted in the foregoing, the focus of optimal control theory is almost always on determination of time dependent functions, rather than purely space dependent design functions. A notable exception is the formulating of Butkovskiy (16), in which a more general problem requiring determination of two control functions is defined, one depending on both time and space variables, the other depending on only one variable, that can be taken as a space variable. On page 44 of Reference (16), he presents a maximum principle characterization of necessary conditions for such problems, in which one of the necessary conditions collapses the time variable through temporal integration, allowing a direct statement of necessary conditions in the design variables, of course involving integrals over time of the state variable. This formulation of necessary conditions has apparently not been exploited, to date. Butkovskiy then proceeds to treat thermal problems, in which the equations of the state and necessary conditions can be obtained in integral equation form. He then obtains analytical and numerical solutions of such problems, with major applications in the fields of thermal systems. While this text is freely referenced in the structural optimization literature, it is not clear that the techniques presented therein, which are based on Soviet results prior to 1965, have been fully exploited.

In a recent review (17) and text (18) Ray and coworkers address distributed parameter optimal control, with emphasis on chemical engineering applications. While the primary emphasis is on dynamic systems, some of the methods may be applicable to distributed parameter structures.

The literature on distributed parameter optimization, since 1970, has taken a turn toward a more abstract formulation of the problem. Lions and coworkers have introduced function space techniques (19, 20), for optimization of linear and nonlinear systems. Their necessary conditions generally result in variational inequalities, a field that has developed quite rapidly during the past decade.

While the applications presented by Lions are rather abstract, the basic ideas employed should have considerable applicability in the field of mechanical system optimization. Full advantage of these techniques has clearly not been taken, to date.

A paper by McGlothin (21) continues the function space setting, but with somewhat less abstraction. A nonlinear dynamic system is treated in a control setting, for which the control variable is time dependent.

The generality of the formulation, however, may allow application to structural problems.

Excellent books on theoretical and computational aspects of optimal control theory include the landmark texts of Pontryagin, et al., (22), Hestenes (23), Bryson and Ho (24), and Cea (25). Pontryagin (22) and Hestenes (23) present very general necessary conditions of optimality for initial-value problems (ordinary differential equations) arising in optimal control theory. Bryson and Ho (24) present iterative numerical methods for direct solution of open loop optimal control problems. Their methods are extended for direct treatment of distributed parameter structural optimization problems by Haug and Arora (13). Cea (25) presents necessary conditions of optimality and numerical methods that may be applicable for distributed parameter optimal design.

A different approach that has been pioneered by Soviet researchers such as Dubovitskii and Milyutin involves an abstract vector space setting, operator theory for representation of the state of the dynamic system, and a variety of formalisms for treating constraints. The abstraction ranges from purely set theoretic, cone ordering techniques from statement of constraints, to functional inequalities and utilization of Frechet differentiation theory. This theory is thoroughly treated in texts by Pshenichnyi (26) and Ioffe and Tihomirov (27). These abstract methods for treatment of distributed parameter optimization problems have not been exploited to date. While the degree of abstraction is high, it appears that operator theoretic properties of continuum systems can be exploited to sharpen the abstract theory and obtain workable necessary conditions and, perhaps more important, a valid mathematical motivation for computational techniques.

Zolezzi (28) published an outstanding paper that is the first known treatment of elliptic state equations in which the coefficients of the differential equation, which is written in divergence form, are allowed to depend upon the control variable. This class of problems represents a valuable model of equations of continuum mechanics and allows treatment of the design problem. The motivation in Zolezzi's paper is in the area of stochastic control, so he also treats parabolic problems, which would be related to heat transfer problems of a continuum. This paper uses functional analysis methods of weak convergence and existence theory of partial differential equations to obtain a set of necessary conditions for the problem of minimizing an integral, subject to the above mentioned differential equations and boundary conditions. The formulation, however, does not include either functional or pointwise inequality constraints.

9.2 BUCKLING OF COLUMNS, PLATES, AND SHELLS

9.2.1 Column Buckling

Historically, one of the first optimal structural design problems addressed was treated by Lagrange in 1770 (29) and later by Clausen in 1849 (30). For an accessible summary of these initial studies, the reader is referred to Reference (31) (Volume I, pp. 66-67 and Volume

II, pp. 325-329, respectively). The first modern treatment of this problem, which sparked substantial interest in optimization by the mechanics community, was presented in an elegant paper by Keller (32). Keller treated the problem of maximizing the fundamental buckling load for a pinned-pinned column, under the condition that the total volume of the column is specified. In this fundamental paper (32), Keller addresses both the question of optimum tapering of the column and selection of the optimum tapering of the column and selection of the optimum cross-sectional geometry. He employs a directional derivative approach to obtain necessary conditions of optimality and obtains closed form solutions. In a subsequent paper, Tadjbakhsh and Keller (33) treat a variety of boundary conditions for column optimization, using the analytical method Keller had earlier presented (28). In later related developments, Farshad and Tadjbakhsh (34) and Gajewski and Zyczkowski (34, 35) treat more general load condition. It is interesting to note that since no lower bound on cross-sectional area or upper bound on stress is specified, zero cross-sections (singularities) occur in the designs obtained. In 1966, Keller and Niordson presented a similar analysis of column optimization, seeking the tallest-column that will remain stable under its own weight (36). The earlier variational approach of References (32) and (33) was employed in this analysis.

In a sequence of papers (37-40), Taylor and Prager develop a variational formulation of the problem of column optimization, employing stationarity of the Rayleigh quotient to obtain optimality criteria for fixed volume and maximum buckling loads, including lower-bounds on cross-sectional areas. In Reference (40), Taylor considers prestress. They prove sufficient conditions of optimality for sandwich columns. In related developments, Huang and Sheu (41) and Frauenthal (42) test optimization of clamped free columns, with a variety of cross-sections, including tubular.

There is a subtle difference in the technical approach to developing necessary conditions of optimality in References (32, 33, 36), where a directional derivative of the buckling eigenvalue with respect to design is calculated, and in References (37-42), where a first variation of the Rayleigh quotient is employed to calculate a Frechet Derivative of the eigenvalue, which is then used with a Lagrange multiplier technique to obtain necessary conditions of optimality. In the second approach, at least a Gateaux derivative of the eigenvalue with respect to design is assumed to exist.

Haug (43) treats the problem of minimum weight design of a clamped-free column with a lower bound on the fundamental buckling load and an upper bound on stress. The problem is formulated and solved analytically using the Pontryagin maximum principle and numerical using a steepest descent method of optimal control theory. An alternate steepest descent numerical method is applied for solution of the same problem in Reference (44) and the numerical solutions presented in References (37-42).

As a result of the zero cross-sections predicted for columns under various support conditions in Reference (32, 33, 36), several authors considered an alternate formulation of the optimum column problem (45-

48), in which the location of the singular points was considered as a design variable. Masur (46) and Olhoff and Taylor determine best location of connections and singularities whereas Farshad (45) and Mroz and Rozvany (47, 48) determine optimum support locations.

In a fundamental paper, Olhoff and Rasmussen (50) show that for a clamped-clamped column, a repeated eigenvalue occurs at the optimum design for certain values of a lower bound on the cross-sectional area. In this paper, the result of Tadjbakhsh and Keller (33) is shown to be in error. They then present a variational formulation of the problem of maximum buckling load, using the first variations of the Rayleigh quotient, with constraints appended using Lagrange multipliers. It is suggested in this paper that the error made by Tadjbakhsh and Keller is one of requiring continuous first derivatives of the buckling mode, whereas at the singular points, that arise the slope must in fact be allowed to be discontinuous. Olhoff and Rasmussen iteratively construct optimal designs for the clamped-clamped column and show that no singular value of the cross-section occurs, even if the lower bound cross-sectional areas is taken as zero. They predict that below a certain value of the lower bound on cross-sectional area, the lower bound constraint will not be active.

Banichuk (51) and Olhoff (52) and coworkers examine the nature of singularities of optimal columns and obtain transversality conditions that must hold at singular points. Komkov and Haug (53) present an argument, using nonlinear analysis of buckling, that indicates that zero cross-sections will generally not occur if a refined optimization formulation is employed. For a detailed technical discussion on column optimization, the reader is referred to Olhoff's paper (54) in these proceedings.

In very recent literature, Masur and Mroz (55-57) have developed an alternate form of optimality criteria for singular problems such as those arising when repeated eigenvalues occur in an optimization problem. They use an optimality criterion that does not allow both eigenvalues to increase in any feasible direction of design modification. Using this necessary condition, they obtain an optimality criterion which reduces to that of Olhoff and Rasmussen (50).

A simplified model of the clamped-clamped column, using discrete torsional springs to represent stiffness of the structure, is employed by Prager and Prager (58) to show that even with a simplified model, repeated eigenvalues occur at certain optimal designs. Prager employs a variable support stiffness that shows that the nature of the optimal design depends on the value of torsional rigidity of the support. A similar problem for the clamped free column, with a flexible support, is treated by Banichuk (59), but in this flexible support configuration no repeated eigenvalues occur.

Haug, Choi, and Rousselet (60-64) use a directional differentiation theory and abstract optimization theory (26) to study structural optimization problems with repeated eigenvalues. They show (61-64) that a repeated eigenvalue is in general not Gateaux differentiable, much less Frechet differentiable. They demonstrate existence of a directional derivative and employ directional derivative ideas with modern optimization theory (26) to obtain a rather complex set of necessary conditions for optimal design problems (60, 61) in which repeated eigenvalues do occur. Several examples are presented in References (60, 61) to show that the clamped-clamped column is not the only structural optimization problem in which repeated eigenvalues occur at an optimum design. Examples are presented in Reference (61) to show that in general, erroneous results may be expected if Lagrange multiplier techniques are used to derive optimality criterion for problems in which repeated eigenvalues occur. Using directional derivative theory developed for repeated eigenvalues in Reference (64) it is shown (61) that for the clamped-clamped column problem treated by Olhoff and Rasmussen (50) and Prager (58), if the design is symmetric above the midpoint of the column and if the first two repeated eigenvalues correspond to symmetric and antisymmetric modes, then the repeated eigenvalues are indeed Frechet differentiable and the variational form of optimality criterion, with Lagrange multipliers, used by Olhoff and Rasmussen (50) can be justified, provided a certain set of eigenvectors corresponding to the repeated eigenvalue are used. As shown by other examples in Reference (61), however, this is not the general case. In fact, if symmetry of the design of the clamped-clamped column is destoryed by other constraints, then the Lagrange multiplier technique may be invalid. These results, supported by the analysis presented by Masur (57) indicate that greater care is required in development of optimality criteria and for the problems in which repeated eigenvalues occur.

As a final note on the column buckling problem, it is important to note that other substantial technical difficulties may arise in design optimization in which repeated eigenvalues occur. It has been shown by Thompson and coworkers (65-68) that imperfection sensitivity may be associated with buckling of structures if repeated eigenvalues occur. They demonstrate that catastrophic failure may result if optimality criterion implying repeated eigenvalues are not adopted. Thus, to assure a safe design, it appears that future effort should be devoted to investigating post-buckling behavior of an optimized design, in which repeated eigenvalues occur. In order to assure imperfection insensitivity for practical designs, additional constraints may be placed on the branching character of the buckling loads in post-buckling behavior. This complex subject will require great technical care and substantial development in optimal design theory. As indicated by the examples presented in Reference (61), occurrence of a repeated eigenvalues at optimum designs is not limited to the clamped-clamped column. Vibration examples are also given that display this character and heuristic optimality criteria discussed by Thomspon and Hunt (68) illustrate that in order to avoid catastrophic nonlinear response when repeated eigenvalues occur, future efforts should be devoted to adding constraints to the design problem to insure that imperfection sensitivity does not occur even in the presence of repeated eigenvalues.

9.2.2 Other Buckling Problems

Only a few papers have appeared in the literature concerning design of plates and shells for buckling behavior. Frauhenthal (69) treated the problem of optimization of an axisymmetric circular plate, subject to axisymmetric radial loading. Under these conditions, the governing diffential equations reduce to ordinary differential equations and problems involving both sandwich and solid cross-section geometries are treated, including stress constraints on the pre-buckled structure. A variational formulation in which the Rayleigh quotient is used to represent the buckling load is employed and constraints are appended with Lagrange multipliers. A somewhat more complicated problem of optimal thickness determination for a cylindrical shell that is loaded by external pressure is treated by Andrev and coworkers (70). A Pontryagin type maximum principle is employed to obtain optimality criterion and iterative algorithm is used to construct numerical solutions. Weight optimization of a reinforced spherical shell under external pressure and the associated buckling constraint is treated by Manevich and Kaganov (71). Approximations are employed to determine local and global buckling loads and elementary optimization methods are employed to construct solutions. Zyczkowski and Kruzelecki (72) formulate shell buckling optimal design problems involving external pressure and bending. They use a variational formulation to obtain optimality criterion and solve two example problems.

Finally, Tadjbakhsh and Farshad (73) formulate an arch optimal design problem in which the arch is presumed to support no bending movement or shear. A variational method is used to formulate optimality criteria and examples exhibiting singular cross-sections are solved.

A limited number of papers treat column buckling in conjunction with other structural performance constraints or other modes of structural deformation. These papers are discussed in Section 9.6.

9.3 VIBRATION OF BARS, BEAMS, AND PLATES

Stimulated by the Keller papers on optimization of columns (32, 33, 36), an intense activity is optimization of structural elements for vibration was initiated by Niordson (74) and Turner (75). In his fundamental paper (74), Niordson uses the directional derivative technique associated with the Rayleigh quotient for vibration of a simply supported beam, to develop necessary conditions of optimality for distribution of a fixed amount of material to maximize fundamental frequency. He uses an iterative technique of the specific example problem, which exhibits singular sections. Turner (75), on the other hand, deals with minimization of total mass of a bar in axial vibration, leading to a second-order differential equation formulation for structural vibration. He uses a discrete element approximation method to construct numerical solutions.

9.3.1 Vibration of Bars

Taylor (76) reformulated the Turner problem of axial vibration of a bar and employed a variational approach to obtain necessary conditions for maximization of the fundamental eigenvalue, subject to a constraint on the volume of material available. The method employed by Taylor is closely related to his development of optimality criterion for column-buckling (37). In a subsequent note, Taylor (77) extends his formulation for axial vibration optimization to include a lower bound on cross-sectional area. He uses the basic variational formulation presented by himself and Prager (39), including lower bound constraints, which is applicable to both buckling and natural frequency problems.

The axial vibration problem was studied by Sheu (78) using piecewise uniform structural elements to construct the rod. He derives optimality criterion and numerically constructs solutions. Sippel and Warner (79) present an extension of the axial vibration problem to a multi-element structure with several bars connected in series to a set of discrete masses. They use the variational formulation of Prager and Taylor (39, 77) to obtain optimality criteria. They construct optimal solutions with variable cross-section within each of the structural elements and subsequently determine an optimum design with constant cross-sections in each of the elements. They conclude that the constant-cross-section of the optimal design is competitive with design obtained by tapering the structural members. Quite recently, Cardou and Warner (80) have developed a more general and rigorous set of necessary conditions of optimality for the axial vibration problem and have applied their method to problems of beam and multi-element structural deformation.

As a final note on the axial vibration problem, Miele and coworkers (81) in a recent paper, apply a gradient projection technique of optimal control theory to numerically construct solutions of the problem of mass minimization of an axial bar, subject to a constraint on natural frequency and cross-section.

9.3.2 Vibration of Beams

Following the initial paper of Niordson on optimal vibration of beams (74), Brach treats the problem of maximization of fundamental frequency for beams in which the moment of inertia of the beam cross-sectional area is proportional to the cross-sectional area. Using a variational formulation, Brach constructs closed form solutions and find that for certain boundary conditions no extremal fundamental frequency may exist. In a note (83) Brach later shows that for this class of structures, the problems of maximization of fundamental frequency with a given amount of material and the problem of minimization of mass with a constraint on natural frequency are not equivalent. The question of existence of solutions of optimization problems with eigenvalue constraints is similarly treated by Vepa (84). In a related paper (85) Brach presents an alternate method of constructing approximately optimum designs, using a characteristic vibration shape technique and studies vibration of beams and bars.

McCar, Haug, and Streeter (86) present a formulation for structural optimization with lower-bound constraints on beam cross-section and on natural frequency, using a gradient projection technique of optimal control theory. They apply the theory to optimization of a portal frame. Since the technique they used does not necessarily control the fundamental frequency, numerical computations encountered difficulty when a shift of fundamental mode from antisymmetric to symmetric occurs. This behavior is now suspected to be similar to the coalessence of eigenvalues encountered in the buckling optimization problem presented by Olhoff and Rasmussen (50). A similar portal frame application of the method of Cardou and Warner (80) was subsequently presented. The numerical method employed in Reference (86) was subsequently refined, using the gradient projection method of Reference (44), to develop a more effective computational technique that is applied to vibration problems in Reference (87). A related numerical treatment of optimization of beams of minimum weight, subject to constraints on natural frequency and a lower bound on cross-sectional area, is presented by Pierson (88). He employs a gradient projection technique of optimal control theory for beam optimization with a variety of boundary conditions. Considerable numerical experience is presented. Kamat and Simitses (89) present a numerical method for natural frequency optimization of a beam on an elastic foundation. They use a finite element discretization and solve the problem using a nonlinear programming method.

The optimality criterion method initially presented by Niordson for beam optimization (74) is extended by Karihaloo and Niordson (90, 91) in treating the problem of maximizing fundamental frequency of beams, with a constraint on the amount of material available, but no lower bound on cross-sectional area. They treat both similar cross-section and cross-section of fixed width and height. Similar results are obtained by Seiranyan (92) for the problems in which no lower bound on cross-section is imposed.

Warner and Vavrick (93) first treat the problem of structural optimization with constraints on several natural frequencies, using a variational formulation. Olhoff (94, 95) addressed the problem of maximizing higher-order eigenfrequencies of beams, first without a lower bound on cross-section (4) and subsequently with lower bounds on cross-sectional area (95). In a related development, Troitskii (96) seeks to maximize the difference between natural frequencies in the eigenspectrum associated with an optimum design. He develops optimality criteria using an axial bar.

Motivated by singular behavior of optimum designs with no lower bound on cross-sectional area, Mroz and Rozvany (97, 98) seek the optimum placement of singularities for extremum eigenvalues. Singularities of optimum beams in vibration are addressed more comprehensively by Olhoff and Niordson (99) and in considerably more detail in Reference (100). They show that there are numerous relationships between the problems of maximizing the nth natural frequency and optimal placement of singularities.

As a final observation on beam optimization for natural frequency, Kamat (101) used a variational formulation of the beam optimization problem to investigate the effect of shear deformation and rotary inertia in optimum designs. In constructing solutions, he employs the finite element computational technique presented in Reference (89).

9.3.3 Torsional Vibration of Shafts

Weisshaar formulates a problem of optimization of a torsional shaft to minimize weight, subject to a condition that the lowest natural frequency be fixed (102). Variable wall thickness of the shaft is to be selected to minimize weight of the structure, subject to the frequency constraint. He employs necessary conditions of optimality given in References (103, 104) and numerically constructs optimum designs. In recent papers, Vavrick and Warner (105, 106) have investigated the problem of minimum mass design with torsional frequency and thickness constraints, in considerable detail. They employ the Pontryagin maximum principle to obtain optimality criteria and construct pieced extremals that lead to optimum designs. They show that, for certain values of upper and lower bounds on thickness, discontinuous optimum designs can arise, even though there is no discontinuous forcing function or any other form of discontinuous input to the problem. They also show (106) that the problem of minimum mass with bounds on eigenvalues and the dual problem of maximization of eigenvalues for a given mass are equivalent.

Minimum weight design problems associated with shafts of high-speed machinery and with turbine disks are treated in References (107-110). Bending dynamics of shaft and turbine blades are incorporated in the optimization problem and bounds on natural frequency are imposed. Optimal control methods and methods of nonlinear programming are employed in obtaining solutions.

9.3.4 Vibration of Plates

The first treatment of variable thickness plate optimization for natural frequency response appears to have been presented by Olhoff (111) in extension of the method of Niordson (74) to an axisymmetric plate for which the eigenvalue problem is an ordinary differential equation. In this paper, Olhoff obtains singular (zero cross-section) plates as optimum. The method of obtaining optimality criterion is the variational method earlier presented by Niordson (74).

From a totally different approach, Armand and Vitte (103) and Weisshaar (104) used optimality criterion of the Pontryagin maximum principle type for development of necessary conditions of optimality for a variety of plate geometries and forcing functions. The primary emphasis of their work was to develop optimality criterion for minimum weight design of structures with constraints on natural frequencies. For the general plate problem, this set of necessary conditions is terribly complex and reduces to a computationally feasible set of necessary conditions only for idealized problems. These results were extended by Armand (112, 113) for treatment of shear panels and a variety of problems arising in aerospace applications. Selected problems are solved in close form, but the complexity of necessary conditions does not lead to a general computational algorithm for more complex problems of plate bending. In a more recent paper (114) Armand presented a numerical method of constructing solutions of variable thickness plate problems for minimum weight, subject to eigenvalue constraints. In this work he shows that there is a tendency for ribs to develop in a plate, to enchance stiffness and increase natural frequency.

Olhoff extended his variational formulation and necessary conditions for the problem of maximizing the fundamental natural frequency of a rectangular plate, subject to the condition that a given volume of material is available for a solid plate (115). In this paper he writes necessary conditions and numerically constructs solutions that vary smoothly and have zero cross-sections appearing. He observes in this paper that these designs should be considered as only local optimum, since one would expect ribs to develop to enhance stiffening of the structure, as noted by Armand (114). Olhoff subsequently treats the problem of singularities and formation of stiffeners in optimal design of plates (116), using optimality criteria. In very recent work, Olhoff and coworkers (117-119) shows that indeed ribs do develop in a natural way if certain upper and lower bounds are placed on plate thickness. It is interesting to note that Vavrick and Warner found the same basic criterion for occurrence of discontinuities in design of torsional vibrating shafts (105).

Lurie (120) shows that the Prager optimality criterion (39) is applicable to plate optimization, but that if one uses this formulation presuming considerable smoothness in the solution of the vibration equations, only smooth designs will occur. He argues that the only way to treat problems with rib stiffeners is to model the structural response with explicit rib stiffeners attached to the variable thickness plate and to determine the optimal location of the stiffener and their geometric characteristics. Seiranyan (121) considers the problem of maximization of fundamental natural frequency for fixed volume of material for an axisymmetric circular plate and finds no ribs.

From a direct gradient direction approach, using function space derivatives of eigenvalues, Haug and coworkers (44, 87, 122) use a gradient projection computational technique to minimize the weight of plates with constraints on natural frequency and cross-sectional thickness. As the cross-sectional lower-bound is decreased, simply supported plates similar to those presented by Olhoff (115) are obtained. Use of the gradient projection algorithm has been compared with results obtained using a discrete finite element formulation of the problem (123) and virtually identical results are obtained.

Virtually no results exist on optimization of shell structures for specified natural frequency. One paper that presents an extension of beam optimization idea to a class of shells of revolution is (124) where optimality criterion are derived but no examples are solved.

9.4 DEFLECTION AND COMPLIANCE OF BEAMS AND PLATES

9.4.1 Beam with Deflection Constraints

In an early paper, Barnett (125) seeks to minimize the weight of a beam, subject to the condition that deflection at a specified point on the beam is bounded. He employs a Castigliano theorem to characterize displacement at the given point and a variational approach to obtain necessary conditions of optimality. Several subsequent papers (126-129) treat the problem of minimum weight design of beams with deflection constraints at specified points. Dixon (126) applies Pontryagin's maximum principle for a cantilevered beam with extreme displacement at its end. Huang and Tang (127) use an extension of the

Barnett technique with the principle of virtual work, applying their necessary conditions to both statically determinant and statically indeterminant beams. Prager (128) employs the principle of stationary mutual potential energy to obtain optimality criterion for beams with given displacement, with displacement governed by applied load and thermoelastic considerations. Finally, Chern (129) treats an extension of the Barnett problem by the same method, but allows the applied load to depend on design.

The problem of minimum weight design of beams with a constraint on the maximum reflection, occurring at an unknown point, was first treated by Haug and Kirmser (130), using the Pontryagin maximum principle to construct solutions by pieced extremals. Application to simply supported and cantilevered beams with fixed width and variable depth are presented, each having a smooth cross-section. The same minimum weight design problem, with constraint on the absolute value of displacement is treated numerically by a gradient projection method in Reference (87) with a load that leads to bicurvature. Results shown there indicate discontinuous cross-sections at the optimum design, a result previously obtained Haug (131), with the more difficult computations arising from the Pontryagin maximum principle.

The same method, employing the Pontryagin maximum principle for optimization of structural elements is treated by DeSilva (132).

Shield and Prager (133) presented a variational formulation of the problem of minimum weight design with constraints on deflection, using a direct variational approach. They prove a principle of stationary mutual potential energy, associated with two kinematically admissible displacement states for the system. One of the displacement states is calculated to be associated with a unit load at the point where displacement is to be constrained. The principle then gives the displacement at this point, under a given load. This variational formulation of the displacement constraint allows the authors to extend the variational method of Prager (39) to the problem of optimization of structures with displacement constraints. Prager and coworkers (134, 135) applied and extended the basic principle developed in Reference (133). They demonstrate that the optimality criterion obtained through the stationary mutual complementary energy approach is sufficient for sandwich beams and necessary for othe beam cross-sections. In a subsequent paper, Huang (136) applies the Shield-Prager technique under a variety of loading conditions. Defalias and Dupuis apply the Shield-Prager method (137-139) using an iterative technique to locate the point at which maximum deflection occurs.

In independent developments, Bhargava and Duffin (140) and Simitses and Kotras (141) develop optimality criterion and solve example problems for cantilevered beams with a constraint on deflection at the end. Bhargava and Duffin (140) employ duality theory and variational techniques to develop a rigorous optimality criterion that is subsequently applied to a cantilevered beam on an elastic foundation, with fixed height and variable width. Simitses and Kotras (141), on the other hand, use dirac-delta functions to represent displacement at the given point and subsequently bound by displacement and seek a minimum weight design. The same problem of minimum weight design of a beam on an elastic foundation, with displacement specified at a known point is treated by Distefano and Todeschini (142), using an invariant imbedding technique.

The problem of minimum weight design of beams with bounds on maximum deflection, occurring at an unknown point on the beam, is treated in References (143-145). Huang (143) employs a Castigliano theorem to represent deflection at a specified point in terms of design and seeks to minimize weight, subject to the condition that the deflection constraint is satisfied at every point along the beam. He treats the point of maximum deflection as a variable in the formulation and determines it as part of the solution of the problem. Komkov and Coleman (144) use a similar argument, but with distribution theory to present deflection at selected points along the beam and subsequently impose the deflection constraint at all points. They develop optimality criterion for both beams and plates. Cinquini (145) uses a direct variational approach to adjoin the deflection constraint to the weight function that is to be minimized. He allows for discontinuities in adjoint variables and develops a set of necessary conditions for the minimum weight deflection problem that resemble those obtained for the same problem by the Pontryagin maximum principle in Reference (130). He then carries out example calculations with statistically determinant beams.

Armand (146) presents a general analysis of optimization of beams with equality and inequality constraints involving deflection and other response measures. Banichuk (147) formulates structural design problems with constraints on deflection over the entire beam as a minimum-maximum problem. He obtains optimality criteria for problems with deflection constraints and for other problems, using necessary conditions of minimum-maximum optimization theory.

9.4.2 Plates with Deflection Constraints

Hegemier and Tang (148) treat the problem of plate optimization with bounds on displacement at several specified points. They employ a variational method to obtain optimality criterion and discretize the plate using finite element calculations to construct numerical solutions. Simply supported and clamped sandwich plate examples are solved. Using a gradient projection function space optimization algorithm, Haug and coworkers (47, 87, 122) solve the problem of minimum weight plate design with constraints on deflection at all points in the plate and other response measures. Armand and Lodier (149) use a finite element formulation for optimization of rectangular plate with a single deflection constraint. In their results they note development of apparent ribs in the design to enhance stiffness. They also discuss the regularity of displacement function that must be allowed in the finite element formulation to enable ribs to form in the iteratively obtained design. In an impressive sequence of papers, Banichuk and coworkers (150-153) develop a theory and computational method for minimum weight design of plates with a deflection constraint over the entire plate. Banichuk first (150) treats the design problem with deflection constraint at a single point and obtains asymptote estimates of the solution. He then (151) uses the minimum-maximum formulation of Reference (147) to obtain optimality estimate of maximum displacement and stress, hence reducing the minimum-maximum problem to an approximate minimum problem. Rectangular plates with a variety of constraints are optimized in Reference (153).

As a final note on plate design with deflection constraints, a paper investigating the influence of rib stiffeners are considered. Simitses (154) has constructed nonoptimum circular plate with ribs and compared its displacement with that of a plate that is optimized for deflection. He finds that the ribbed design yields smaller deflection, so the nonribbed optimum must be only a local optima.

9.4.3 Beams and Plates with Compliance Constraints

Work done by the applied load system in undergoing displacement (compliance) is often treated as a measure of stiffness of the structure. This compliance measure is attractive from an optimization point of view, since it is a global measure of structural performance; i.e., it can be written as an integral over the entire element.

Prager and coworkers made extensive use of compliance constraints (155-158) in formulating and obtaining optimality criterion for numerous structural optimization problems. They obtain optimality criterion using the basic method in Prager's landmark paper with Taylor (9) and specifically treat problems of minimum weight with piecewise constant section properties (155), beams and frames (155, 158), and Michell trusses (157). Huang (159, 160) uses the same basic approach for minimum weight design of beams and circular plates. A similar set of problems is addressed by Chern and Prager (161-161b) using optimality criteria associated with the compliance constraint. Save (162) formulates a more general problem, including many design variables and employs compliance constraints to reflect stiffness and deflection of the structure.

Masur, in a sequence of papers (163-163b) develops a consistent optimization theory for optimization of beam and plate structures for compliance. He first (163) develops a rigorous set of optimality criteria for material distribution for minimum volume, with constraints on compliance. He then (163a) treats the practical problem of selecting among available structural sections to optimize design with compliance constraints.

Finally (163b), he carefully studied the problem of maximizing stiffness (compliance), specifically studying the problem of a circular plate with a uniform pressure. He derives optimality criterion and shows that hinges occur in the optimum design. He then analyzes the effect of discontinuities associated with discreteness and shows that irregularities may arise in the optimum design, even though no irregularity on the load was specified. Masur expands upon considerations of singularity and nonstationarity of optimal designs in Reference (54).

Mroz and Rozvany (164, 165) treat optimization of a variety of structural elements with compliance and other global performance constraints. Mroz (164) shows that in some problems global theorems may be obtained, whereas in others only local conditions are possible. Mroz and Rozvany (165) then seek to determine optimum location of supports with constraints on compliance. Reiss (166) also develops optimality criterion for maximum compliance associated with an axisymmetric plate.

9.4.4 Miscellaneous Problems

Chern and Prager (167) formulate a problem of optimal thickness of an axisymmetric rotating disk with a bound on displacement of the edge. They use the Shield-Prager (133) principle of stationary mutual potential energy to derive necessary and sufficient conditions of optimality and construct numerical solutions. Chern (168) treats a related problem of a rod spinning about a line perpendicular to its axis, with thermal effects and a constraint on deflection at the end.

Finally, Fuchs and Brull (169) formulate a problem of optimizing support location for a beam, with an objective of minimizing the maximum deflection. They then use a strain energy approximation for the minimum-maximum deflection problem and solve examples.

9.5 DYNAMIC RESPONSE

9.5.1 Froced Steady State Oscillation

As an extension of structural optimization with bounds on natural frequencies, several authors have investigated design optimization of structures with harmonic steady-state input. Icerman (170) formulates a problem of minimizing virtual work of the applied load for steady-state forced oscillation of rods, beams, and trusses. He uses a variational formulation to obtain optimality criterion, from which he obtains analytical solutions for axial motion of a rod bending of a beam, and response of a truss. Mroz (171) formulates a general problem of structural dynamic optimization under harmonic loads. He also uses a dynamic compliance, which is the work done by the applied forces during one cycle of motion, as his optimality criterion and develops a variational formulation of the problem and optimality criterion. His variational necessary condition is applicable to broad classes of structures and is proved to be both necessary and sufficient for sandwich structures. He applied the theory to sandwich and solid beams and plates.

Plaut (172) formulates the problem of minimum weight design of structures under periodic loading, subject to the constraint that deflection of a specified point of the structure is prescribed. He extends the principle of stationary mutual potential energy of Shield and Prager (133) and derives optimality criterion for elastic sandwich beams. Plaut subsequently (173) uses a Rayleigh-Ritz technique to approximate optimal design of structures for dynamic response.

Huang (174) expands on the periodic loading problem treated by Plaut (172), to include the effect of inertial force. He also employs the principle of stationary mutual potential energy and obtains optimality criterion. The development is applicable to general elastic structural systems. Rockenback (175) formulates a generalization of the problem treated by Plaut (172), to include deflection constraints under harmonic exitation for a specified range of exitation frequencies. He treats the problem as a nonlinear programming problem, using penalty function techniques to construct numerical solutions.

A related class of optimal design problems involves selecting of support parameters to optimize dynamic response of a structure that is subjected to harmonic or statistically defined ground input motion. Typical treatments of such isolation problems may be found in References (176-178). For a detailed review of literature regarding optimization of support substructures for dynamic isolation, References (179, 180) may be consulted.

9.5.2 Transient Dynamic Response

Brach (181, 182) appears to be the first to address transient response optimization of beams with suddenly applied load, seeking to minimize an integral measure of impulse response, subject to the constraint that a fixed amount of material is available. A gradient projection numerical method is employed to numerically solve the problem, after a discrete model of the beam is constructed.

Plaut (183) formulates a minimum weight beam design problem, under general distributed transient load and subject to an upper-bound constraint on the deflection of the beam. He derives upper-bounds for displacement response to input and finds a design to minimize weight, subject to the condition that the upper bound does not exceed the deflection limit. He notes that the question of precision of the upper bound is unanswered and that conservative designs may be fact be generated. He gives applications to cantilever beam design optimization. A problem of similar level of generality is formulated by Pochtman (184), who seeks to minimize weight, subject to constraints on maximum displacement response over time and domain of the plate being considered. He treats general transient load and incorporates constraints on natural frequency. The design of the plate is characterized by a finite number of design parameters and a random search technique is used to numerically construct solutions.

In the first general computational approach to optimization of structures subject to transient dynamic response, Fox and Kapoor (185) present a technique based on an upper-bounding idea similar to that employed by Plaut (183). Fox and Kapoor use a finite element shock spectrum response estimate that gives upper bounds onpeak displacement to shock loading. These bounds are stated in terms of eigenvalues and eigenfunctions of the finite element matrices, which are differentiated to obtain gradient information needed for feasible direction optimization. Two numerical examples involving truss-frame structures are presented. The significance of this paper should not be underestimated, since it represents the first step in development of a general approach to structural optimization with transient dynamic response constraints.

After a lapse of 6 years following the Fox and Kapoor paper (185), two papers presenting substantially more general approaches to the problem appeared almost simultaneously (186, 187). Cassis and Schmit (186) seek to minimize weight, subject to constraints on stress and displacement at modes within the finite element model of the structure. They employ approximation concepts to obtain expressions for derivatives of peak displacement and stress response with respect to design. They then employ an exterior penalty function, sequentially unconstrained minimization technique for numerical solution of design problems. They present examples involving frame structures with from 3 to 21 members. They also generate plots of feasible design regions for simplified structures that illustrate very complex constraint boundaries and, in some cases, disconnected feasible regions. Using an equivalent integral functions constraint formulation for inequality constraints on displacement and stress within the structure, Feng, Arora, and Haug (187) develop a formulation for optimization of structures under general transient dynamic response. They use a gradient projection technique and a finite element numerical analysis method for optimization of structures. No approximations are made in modeling of the structure, other than selecting the number of modes to be employed in eigenfunction expansion of the dynamic solution. Example problems involving up to seven truss-frame elements are solved and numerical results presented. In subsequent discussion (188) Feng, Arora, and Haug treat example problems presented in Reference (186), using the gradient projection technique of Reference (187). It is shown (188) that results obtained with the sequential unconstrained minimization technique of Reference (186) are at best local optima and substantially conservative.

The gradient projection technique employed in Reference (187) was extended to a distributed parameter function space formulation (189, 190) and used to solve a variety of design optimization problems, including simply supported beams, clamped beams, cantilevered beams, simply supported plates, and clamped plates. The numerical algorithm is based on perturbation theory of differential operators and uses an adjoint variable method for design derivative calculation. Finite element methods were used for numerical analysis and the iterative optimization method tested for convergence. Numerical results obtained with the continuous formulation (189, 190) compare quite favorably to the parallel approach in a finite development design space setting (187).

Generalizations of the methods of References (187, 189, 190) are presented in References (191, 192) with applications to structural and machine elements, vibration isolators, and vehicle suspension systems. A more rigorous and generally applicable development of distributed parameter structural optimization for dynamic response is presented, with applications, in Reference (193), based on design sensitivity analysis methods for operator equations of mechanics of References (63, 64) and the gradient projection method of Reference (122).

9.5.3 Earthquake Structural Design

Kato, Nakamura, and Anraku (194) first addressed the explicit problem of optimization of shear building structures for earthquake resistance. They use ground motion as an exitation to a linear structural model with damping and predict dynamic response of each floor of the structure. Linear approximations are then made to obtain design derivatives of dynamic response and sequential linear programming is employed to optimize the design. Venkayya and Khot (195) present a technique for optimization of structures due to impulsive-type loading encountered in earthquake situations, as well as in aircraft structural design. They seek minimum weight with bounds on dynamic stiffness of the structure, defined by the Rayleigh quotient of the equations of motion, and bounds on stress. They use a displacement finite element model of the structure and develop an optimality criterion, using a Lagrange multiplier formulation, that implies a constant value of an energy measure throughout the structure if the structure responds in a single mode. They then extend this to an approximate technique for structures responding in a combination of modes. An iterative redesign algorithm is presented, based on constant distribution of their energy optimality criterion. The present applications of the technique to a wing span of an aircraft, modeled with 155 bar elements, and to a circular arch.

Cheng and Botkin (196) formulate a general problem of structural optimization for dynamic response using model analysis, obtaining an upper-bound for displacement and stress response. They differentiate through the equations of motion with respect to design parameters to obtain derivatives of the response measures with respect to design and employ a feasible direction algorithm for optimization. They present several examples of frame structural optimization associated with earthquake-type loading. In subsequent papers, Cheng and Srifuengfung (197, 198) adopt the optimality criterion method presented by Venkayya (195) and calculate optimal designs of frame structures.

In 1974, Ray, Pister, and coworkers (199-201) treated the earthquake structural design problem using linear structural models and optimizing on structural reliability. Feasible direction optimization methods are employed for selection of design parameters, under failure constraints associated with earthquake ground motion. In a more fundamental development, Ray, Pister, and Polak (202, 203) developed a method for calculating derivatives of transient dynamic response of structures to earthquake inputs. Nonlinear behavior of structural response is incorporated in their analysis for multistory frame structures. They present a numerical integration method for calculating explicit design derivatives of dynamic response, as needed by optimization algorithms. A specified earthquake isolation system for building structures is optimized by Bhatti, Pister, and Polak (204) for dynamic response constraints, using a feasible direction optimization method. Subsequently (205), they published a general-purpose optimization program capable of treating such problems using a reliable general purpose optimization program. Bhatti (206) presents a detailed treatment of optimization of an earthquake energy absorbing device at the foundation of a tall building. The building itself is modeled by elastic structural response, and the isolator is allowed to undergo plastic deformation, under earthquake shock loading. Performance constraints under small earthquakes, requiring elastic response, and under larger earthquakes in which the structure is allowed to go plastic. A minimum-maximum optimization formulation is employed.

In (207-210) Pister, Polak, and Bhatti present a comprehensive treatment of dynamic structural optimization for earthquake applications.

9.5.4 Miscellaneous Problems

Levy and Wolf (211) investigate the existence of a fully stressed design for dynamically loaded structures and show that such a design does exist for some statically determinate structures.

Komkov and Coleman (212, 213) investigate an optimal control approach to design sensitivity analysis and optimization of dynamically loaded structures. They investigate, from an analytical point of view, optimal damping of structural response.

Therman (214) treats a variety of dynamic structural optimization problems using the Pontryagin maximum principle. He treats problems with constraints on one and two natural frequencies and on response to harmonically varying input.

9.6 SHAPE OPTIMAL DESIGN

9.6.1 Numerical Methods for Shape Optimization

One of the first treatments of the general problem of selection of shape of a structure as the design variable is presented by Zienkiewicz and Campbell (215). The formulate the shape optimal design problem using a finite element model of complex structures and treat the location of the nodal points of the finite element model as design variables. They then calculate derivatives of stiffness and load matrices with respect to design parameters and obtain derivatives of structural response measures and employ sequential linear programming for numerical solution. They present examples associated with dams and rotating turbine machiners. Ramakrishnan and Francavilla (216) employ a similar finite element formulation, but they use a penalty function method for numerical optimization. Francavilla, Ramakrishnan, and Zienkiewicz (217) employ the finite element method of References (215, 216) for fillet optimization to minimize stress concentration.

More basic approaches for surface contouring to minimize stress concentration were initiated by Tvergaard in selecting the optimum shape of a fillet (218). He employs a stress field model of the fillet, with a finite dimensional family of perturbations allowed in the boundary shape, defined in terms of coordinate parameters. He employs a variational analysis of the stress field equations to obtain derivatives of stress with respect to his parameters and uses sequential linear programming to iteratively construct an optimum design. Kristensen and Madsen (219) formulate a class of shape optimal design problems for planar solids, which generalizes the approach presented by Tevrgaard (218). They use orthogonal polynomials to locate the boundary of the body and treat the coefficients in these polynomials as design parameters. They employ a finite element model of the structural response to obtain derivatives of stress with respect to their design parameters and employ sequential linear programming to solve the optimization problem. They solve an elementary problem of the optimum shape of a hole in a biaxial stress field anlytically. They numerically illustrate the method on more complex problems. Bhavikatti and Ramakishnan (22) present a refinement of the formulation of References (215, 216, 217) for optimum design of fillets in flat and round tension bars. They also use a polynomial with coefficients taken as the design variables to characterize the shape of the fillet and a finite element model to calculate stress with the body. They investigate minimization of stress concentration factor, minimum volume design, and design for uniform stress distribution along the fillet boundary as optimality criterion. Derivatives of response measures with respect to design parameters are calculated using a finite element model and sequential linear programming is employed for numerical solution.

A function space gradient projection method of optimal design of the shape of two-dimensional elastic bodies has recently been presented by Chun and Haug (221, 222), using design sensitivity analysis methods similar to those presented by Rousselet and Haug (223) and a gradient projection method of the kind presented in Reference (122). The design objective in this work is weight minimization, with constraints on Von Mises yield stress and shear stress distribution on the boundary.

9.6.2   Shape of Cross-Section of Shafts in Torsion

The problem of optimization of cross-sectional shape of torsion members was first addressed by Henry (224). He develops an analytical method for location of the boundary in terms of a small number of parameters and iteratively selects these parameters to minimize weight, subject to constraints on cross-sectional geometry and boundary stress. He treats the problem of a shaft with grooves required for keyways.

Banichuk (225, 226) formulates a general problem of selecting the optimum shape of cross-section for a nonhomogeneous shaft to maximize torsional stiffness, with a given amount of material available. He uses the fact that the functional minimized by the warping function in a variational formulation of the boundary-value problem is proportional to the torsional stiffness of the shaft. He then takes variations of this functional with respect to both the warping function and boundary variation, using the material derivative idea of continuum mechanics, and obtains a necessary condition for optimum location of the boundary. He treats both simply connected and multiply connected cross-sections. Kurshin and Onoprienko (227) treat the same problem of maximum torsional stiffness of a shaft with doubly connected cross-section, using a complex variable method to determine the optimum boundary.

Banichuk (228) subsequently presents an extension of the torsional stiffness maximization problem for rods, using optimal distribution on a given amount of stiffening material around the boundary. The method he employs is a direct extension of that used in References (225, 226). Gurvitch (229) presents an alternate analytical technique for optimizing the shape of an interior boundary that is associated with inhomogeneity in material, using a coordinate system associated with the warping function and obtaining necessary and sufficient conditions of optimality.

9.6.3   Shapes of Holes in Planar Solids

Neuber (230, 231) and Cherpanov (232) treat the problem of finding a hole shape in a planar solid so as to make tangential normal stress acting on the boundary constant, under the assumption that this is a condition of optimality. Cherpanov (232) cites considerable earlier Soviet literature addressing the same design objective. Wheeler (233) investigates conditions under which constant tangential normal stress along the boundary of a hole, fillet, or notch in a valid optimality criterion for minimum peak stress. He develops criterion for axisymmetric torsion problems and plane problems involving holes, notches, and fillets.

Banichuk (234) formulates the problem of selecting hole shape in an infinite plane body that is in biaxial tension at infinity to minimize peak stress in the vicinity of the hole. Using the maximum principle for harmonic functions, he proves that the optimum hole shape leads to constant tangential normal stress around the boundary of the hole, hence proving the optimality criterion employed in References (230-232) for this class of problems. In a related paper (235) Banichuk treats the problem of finding the optimum hole shape to minimize the maximum value of the second invariant of the stress tensor deviator over an entire plate, which is subjected to uniform bending at infinity. Again using the maximum principle for harmonic functions, and an intricate argument, he shows that the maximum stress occurs at the hole boundary and that for certain classes of problems the hole boundary is uniformly stressed.

9.6.4   Miscellaneous Shape Optimal Design Problems

Banichuk and Karihaloo (236) seek the shape of the cross-section in a cylindrical bar to minimize weight, subject to constraints on torsional stiffness and bending stiffness. They use variational formulation of the torsion problem and a Lagrange multiplier technique to adjoin constraints to the cost function. Using a material derivative type calculation, the first variation of the augmented cost function is taken with respect to shape and optimality criterion are derived. Parbery and Karihaloo (237) use the same method to optimize hollow cylinders, with constraints on torsional and bending stiffness.

Cherkaev (238) presents a theoretical treatment of the problem of boundary shape selection to minimize volume of the structure, subject to a lower bound constraint on natural frequency. He develops a general necessary condition and shows that the variational formulation of Prager (135) can be applied to obtain the same result. As a final note, Durelli and Rajaiah (239) present an experimental method using photo-elasticity to find the optimum shape of a hole in a flat plate, under uniaxial load, to minimize stress concentration.

9.6.5 Related Literature on Domain Optimization

Cea, Zolesio, and Rousselet (240-247) present techniques and applications, from fields other than structural optimization, for selecting optimum domain. They cite substantial literature in this general field and give examples, in diverse engineering disciplines, that hold potential for optimality criteria and direct numerical methods for optimization of structures for shape.

9.7 MULTIPURPOSE STRUCTURES AND MISCELLANEOUS OPTIMAL DESIGN PROBLEMS

9.7.1 Multipurpose Structurs

In a landmark paper (248) Prager and Shield formulate the problem of minimum weight design of sandwich structures, subject to several constraints. They treat a beam-tie, as an example, for specified traverse and longitudinal stiffness. A Lagrange multiplier technique is employed to obtain necessary conditions, which are solved for a simple problem. They discuss an extension of the technique to incorporate other constraints, such as buckling load, that can be represented by a global measure of structural response. Martin and Chern (249, 250) employ the method of Reference (248) to develop optimality criterion for a structure under multiple loads and a bound on minimum cross-section. They treat both continuous and piecewise uniform cross-sections and construct solutions, using optimality criterion with Lagrange multipliers.

Karihaloo and Niordson (251) develop optimality criterion to maximize fundamental frequency of a vibrating beam, under constraints that the volume of material is fixed and that the buckling load is bounded from below. They use Lagrange multipliers to develop an optimality criterion and extend the earlier work of Niordson (74).

Sherman and Wang (252, 253) treat the problem of volume minimization of an axisymmetric plate of varying thickness, subject to constraints that stress and deflection are bounded. They allow exponentially varying plate thickness, hence reducing the optimization problem to one of selecting finite number of parameters through a nonlinear programming approach.

In a sequence of papers (254-257) Karihaloo, Parbery and Wood treat multipurpose tie-beams and beam-columns that are quite similar to problems treated in Reference (248). They use Lagrange multipliers and a variational formulation of the constraints to obtain optimality criterion and present numerical solutions of test problems.

Haug and coworkers (44, 87, 122) use a function space gradient projection optimization technique for minimization of weight of beam, plate, and composite structures under a combined set of constraints on deflection, stress, natural frequency, and buckling loads. The gradients required in the gradient projection calculation are computed using adjoint design variable sensitivity methods of References (63, 64, 258, 259).

Multiple failure criteria constraints, under the same loading system are treated by several authors. Haug and Kirmser (130) use necessary conditions from optimal control theory to obtain pieced extremals for minimum weight design of a beam under both stress and displacement conditions. Similarly, Huang (260) extends the method of Prager and Taylor (39) to incorporate the effect of shear deformation in optimization of beams and columns for buckling and vibration.

In an important sequence of papers (261-264), Seiranyan and Gura analyze minimum weight design of beams and plates with multiple loading and multiple constraints, primarily vibration and frequency. They employ a Lagrange multiplier method for rigorous development of optimality criterion and carry out explicit solution of simple problems. In Reference (263), they use a quasi-optimal technique for approximating solutions of more complex problems. A distinguishing character of this work, as compared to much of the preceding work, is that it employs modern functional analysis differentiation techniques and abstract optimization theory in development of optimality criterion. In an important subsequent paper (265), Seiranyan rigorously demonstrates that most of the functional arising in structural optimization, involving natural frequency, buckling, displacement, and compliance are homogeneous functionals. He employs functional analysis differentiation theory to rigorously obtain optimality criterion for multiple constraint types and gives a clear indication of a unifying direction that may be taken in future efforts and distributed parameter optimization. The method he employs is closely related to abstract optimization theory employed by Choi, Haug, and Rousselet (61) in developing rigorous optimality criterion for general structural optimization problems involving repeated eigenvalues.

There appears to be great potential for organized development of optimality criterion for very general problems, using the methods of References (61, 265). This feeling is strengthened by an approach concerning existence and stability of solution of optimal design problems by Velte and Villaggio (266). They employ modern Sobolev space theory, which has become the foundation of modern theory of partial differential equations, to carefully analyze the existence, uniqueness, and stability of problems of structural optimization. Their study of optimization of a rod shows that under relatively weak hypotheses, existence is guaranteed, but uniqueness and stability are problem dependent.

9.7.2 Special Problems

A problem of optimal determination of distribution of shear modulus in a nonhomogeneous material that makes up an elastic bar in torsion is investigated by Klosowicz and Luri (267). Variation in shear modulus is taken as the design variable and the objective to maximize torsional rigidity, subject to the constraint that a given amount of material is available. They employ a variational method to obtain optimality criterion and discuss algorithms to obtain numerical solutions, but present no numerical results. Banichuk (268, 269) treats a related problem of optimal anisotrophy of rods in torsion and subsequently (270) presents a more general formulation of optimal orientation of axes of anisotrophy in plane deformable solids. He formulates optimality criterion in which material properties play the role of design functions. Hirano (271) treats a similar problem, seeking optimum fiber orientation in multilayer composite materials, to maximize buckling load of a structure.

Banichuk (272, 273) has also treated an interesting class of problems, in which he seeks to optimize some pointwise measure of structural performance, subject to the condition that constraints hold for an entire family of load systems. This problem (272) may be viewed as one with maximization over the load systems in the constraint. Alternatively, (273), he treats problems in which the functional to be minimized is the maximum value of a functional over all load classes. In this way, minimum-maximum problems associated with game theory arise. Optimality criterion or such problems are discussed, but solution methods are extremely difficult.

Aristov and Troitskii (274) address the problem of minimum weight design of beams and plates subject to stress constraints. They formulate the problem as a first order system of differential equations and apply necessary conditions of optimality from optimal control theory to obtain necessary conditions for the structural problem. They also account for discontinuities that may arise in the functional being treated.

Samsonov (275) treats the problem of optimum location of rib stiffeners in plates, with multiple failure criteria. The approach used is similar to shape optimal design, since the rib separates subdomains in which plate action occurs.

Soldovnikov (276) discusses the general problem of determining the shape of generating curves for shells of revolution to minimize weight, subject to constraints on stress and stability. He employs the shell equations to determine stress derivative and sequential linear programming to solve the optimization problem.

Olhoff and Taylor (277, 278) have formulated a class of structural optimization problems called "structural remodeling." Such problems include most classical cost and constraint functions. The idea is to allow use of only a specified amount of material to vary, or remodel, the design to provide the greatest decrease in cost.

Batterman and Pavicic (279) treat minimum weight design of laminated, anisotropic shells of revolution, with shell thickness as the design variable. Loading, fiber alignment, and material properties are specified. A membrane model of shell deformation is employed and constraints on stress are imposed in minimizing weight.

Taylor (280) and Rozvany (281) present generalizations of optimality criterion developed earlier by Prager and Taylor (39) and show that extended version of optimality criterion can often include new problems and models treated in earlier literature.

The design problem of elastic body contour shape selection to minimize peak contact stress was first treated by Conry and Seireg (282). They treat gear contour design using a nonlinear programming method. Haug and Kwak (283) treat a somewhat more general formulation, in which bounds on the allowable position of the surface are included. They use a sequential linear programming method for solution. Lukasiewicz (284) has addressed a similar problem, but uses constant contact stress as an optimality criteria and constructs solutions using the Hertz formulas for contact stress. Very recently, Benedict (285) has developed a rigorous optimality criteria for contour optimization to minimize contact stress for general elastic bodies in contact.

9.8 REFERENCES

1. Wasiutynski, A. and Brandt, A., "The Present State of Knowledge in the Field of Optimum Design of Structures," Applied Mechanics Review, Vol. 16, No. 6, 2963, pp. 341-350.

2. Sheu, C. Y. and Prager, W., "Recent Developments in Optimal Structural Design," Applied Mechanics Review, Vol. 21, No. 10, 1968, pp. 985-992.

3. Pierson, B. L., "A Survey of Optimal Structural Design Under Dynamic Constraints," International Journal of Numerical Methods in Engineering, Vol. 4, 1972, pp. 491-499.

4. Niordson, F. I. and Pedersen, P., "A Review of Optimal Structural Design," Proceedings, 13th International Congress on Theoretical and Applied Mechanics, Springer-Verlag, 1973.

5. McIntosh, S. C., "Structural Optimization via Optimal Control Techniques," Proceedings, ASME Structural Optimization Symposium, ASME, AMDT, New York, 1974.

6. Olhoff, N., "A Survey of the Optimal Design of Vibrating Structural Elements, Part I: Theory" and "Part II: Applications," Shock and Vibration Digest, Vol. 8, No. 8, pp. 3-10, and No. 9, pp. 3-10.

7. Rangacharyulu, M. A. V. and Done, G. T. S., "A Survey of Structural Optimization Under Dynamic Constraints," Shock and Vibration Digest, Vol. 11, No. 12, 1979, pp. 15-25.

8. Venkayya, V. B., "Structural Optimization: A Review and Some Recommendations," International Journal of Numerical Methods in Engineering, Vol. 13, No. 2, 1978, pp. 203-228.

9. Weisshaar, T. A. and Plaut, R. H., "Structural Optimization Under Nonconservative Loading," 1980, see Reference 286.

10. Prager, W., Introduction to Structural Optimization, Courses and Lectures: International Centre for Mechanical Science, Vdine, No. 212, Springer-Verlag, Vienna, 1974.

11. Sawczuk, A. and Mroz, Z., Optimization in Structural Design, Spinger-Verlag, New York, 1975.

12. Rozvany, G. In. N., Optimal Design of Flexural Systems, Pergamon Press, New York, 1976.

13. Haug, E. J. and Arora, J. S., Applied Optimal Design, Wiley-Interscience, New York, 1979.

14. Robinson, A. C., "A Survey of Optimal Control of Distributed Parameter Systems," Automatica, Vol. 7, No. 3, 1971, pp. 371-288.

15. Lurie, K. A., "The Mayer-Bolza Problem for Multiple Integrals: Some Optimum Problems for Elliptic Differential Equations Arising in Magnetohydrodynamics," in *Topics in Optimization*, (Ed. Leitmann), Academic Press, New York, 1967, pp. 147-197.

16. Butkovskiy, A. G., *Distributed Control System*. American Elsevier Publishing Co., New York, 1969.

17. Ray, W. H., "Some Recent Applications of Distributed Parameter Systems Theory--A Survey," *Automatica*, Vol. 14, 1978, pp. 281-287.

18. Ray, W. H. and Lainiotis, D. G., *Distributed Parameter Systems*, Marcel Dekker, New York, 1978.

19. Lions, J. L., *Optimal Control of Systems Governed by Partial Differential Equations*, Springer-Verlag, New York, 1971.

20. Lions, J. L., *Some Aspects of the Optimal Control of Distributed Parameter Systems*, SIAM Series in Regional Conferences in Applied Mathematics, 1972.

21. McGlothin, G. E., "Optimal Control of Distributed Parameter Systems with Penalties on Special Derivatives of the State," *International Journal of Control*, Vol. 24, No. 4, 1976, pp. 145-166.

22. Pontryagin, L. S., Boltyanskii, V. G., Gamkrelidze, R. V., and Mishchenko, E. F., "The Mathematical Theory of Optimal Process," Wiley-Interscience, New York, 1962.

23. Hestenes, M. R., *Calculus of Variations and Optimal Control Theory*, Wiley, New York, 1966.

24. Bryson, A. E. and Ho, Y. C., *Applied Optimal Control*, Wiley, New York, 1979.

25. Cea, J., *Optimization-Theory and Algorithms*, Tata Institute, Bombay, 1978.

26. Pshenichnyi, B. N., *Necessary Conditions for an Extremum*, Marcel Dekker, New York, 1971.

27. Ioffe, A. D. and Tihomirov, V. M., *Theory of Extremal Problems*, North-Holland, Amsterdam, 1979.

28. Zolezzi, T., "Necessary Conditions for Optimal Controls of Elliptic or Parabolic Problems," *SIAM Journal of Control*, Vol. 10, No. 4, 1972, pp. 594-607.

29. Lagrange, J. L., "Sur la Figure des Colonnes," *Miscellanea Taurinensia V*, 1770-1773, p. 123.

30. Clausen, T., "Uber die Form Architecktonischer Saulen," *Melanges Mathematiques et Astronomiques I*, 1849-1855, p. 279-294.

31. Todhunter, I. and Pearson, K., A History of the Theory of Elasticity, Dover, New York, 1960.

32. Keller, J. B., "The Shape of the Strongest Column," Archives of Rational Mechanics and Analysis, Vol. 5, 1960, pp. 275-285.

33. Tadjbakhsh, I. and Keller, J. B., "Strongest Columns and Isoperimetric Inequalities for Eigenvalues," Journal of Applied Mechanics, Vol. 9, 1962, pp. 159-164.

34. Farshad, M. and Tadjbakhsh, I., "Optimum Shape of Columns with General Conservative End Loading," Journal of Optimization Theory and Applications, Vol. 11, No. 4, 1973, pp. 413-420.

35. Gajewski, A. and Zyczkowski, M., "Optimal Design of Elastic Columns Subjected to the General Conservative Behavior of Loading," ZAMP, Vol. 21, No. 52, 1971, pp. 806-818.

36. Keller, J. B. and Niordson, F. I., "The Tallest Column," Journal Mathematics and Mechanics, Vol. 34, 1967, p. 486.

37. Taylor, J. E., "The Strongest Column: An Energy Approach," Journal of Applied Mechanics, Vol. 34, 1967, p. 486.

38. Taylor, J. E. and Liu, C. Y., "Optimal Design of Columns," AIAA Journal, Vol. 6, 1968, pp. 1496-1502.

39. Prager, W. and Taylor, J. E., "Problems of Optimal Structural Design," Journal of Applied Mechanics, Vol. 35, 1968, pp. 102-106.

40. Taylor, J. E., "Optimal Prestress Against Buckling: An Energy Approach," International Journal of Solids and Structures, Vol. 7, No. 2, 1971, pp. 213-223.

41. Huang, N. C. and Sheu, C. Y., "Optimal Design of an Elastic Column of Thin Walled Cross Section," Journal of Applied Mechanics, Vol. 35, No. 2, 1968, pp. 285-288.

42. Frauenthal, J. C., "Constrained Optimal Design of Columns Against Buckling," Structural Mechanics, Vol. 1, 1972, pp. 79-89.

43. Haug, E. J., "Two Methods of Optimal Structural Design," Developments in Mechanics, Vol. 5, Ed., Weiss, H. J., Young, D. F., Riley, W. F., and Rogge, T. R., Iowa State University Press, 1969, pp. 847-860.

44. Haug, E. J., Pan, K. E., and Streeter, T. D., "A Computational Method for Optimal Structural Design II: Continuous Problems," Numerical Methods in Engineering, Vol. 9, 1975, pp. 649-667.

45. Farshad, M., "Optimum Shape of Continuous Columns," International Journal of Mechanical Science, Vol. 16, No. 8, 1974, pp. 597-602.

46. Masur, E. F., "Optimal Placement of Available Sections in Structural Eigenvalue Problems," *Journal of Optimization Theory and Applications*, Vol. 15, 1975, pp. 85-101.

47. Mroz, Z. and Rozvany, G. I. N., "Optimal Design of Structures with Variable Support Conditions," *Optimization Theory Applications*, Vol. 15, 1975, pp. 85-101.

48. Mroz, Z. and Rozvany, G. I. N., "Column Design: Optimization of Support Conditions and Segmentation," *Journal of Structural Mechanics*, 1977, pp. 279-290.

49. Olhoff, N. and Taylor, J. E., "Designing Continuous Columns for Minimum Total Cost of Material and Interior Supports," *Journal of Structural Mechanics*, Vol. 6, 1978, pp. 367-382.

50. Olhoff, N. and Rasmussen, S. H., "On Single and Bimodal Optimum Buckling Loads of Clamped Columns," *International Journal of Solids Structures*, Vol. 13, 1977, pp. 605-614.

51. Banichuk, N. V. and Karihaloo, B. L., "On the Solution of Optimization Problems with Singularities," *International Journal of Solids Structures*, Vol. 13, 1977, pp. 725-733.

52. Olhoff, N. and Niordson, F. I., "Some Problems Concerning Singularities of Optimal Beams and Columns," *ZAMM*, Vol. 59, 1979, pp. T16-T26.

53. Komkov, V. and Haug, E. J., "On the Optimum Shape of Columns," 1980, see Reference 286.

54. Olhoff, N., "Optimization of Columns Against Buckling," 1980, see Reference 286.

55. Masur, E. F. and Mroz, Z., "On Non-Stationary Optimality Conditions in Structural Design," *International Journal of Solids Structures*, Vol. 15, 1979, pp. 503-512.

56. Masur, E. F. and Mroz, Z., "Singular Solutions in Structural Optimization Problems," in *Proc. IUTAM Symposium on Variational Methods in the Mechanics of Solids*, (Ed. S. Nemat-Nasser), Northwestern University, Evanston, Illinois, USA, 1978, Pergamon Press (to appear).

57. Masur, E. F., "Singular Problems of Optimal Design," 1980, see Reference 286.

58. Prager, S. and Prager W., "A Note on Optimal Design of Columns," *International Journal of Mechanical Science*, Vol. 21, 1979, pp. 249-251.

59. Banichuk, N. V., "Optimizing the Stability of a Bar with Elastic Clamping," MTT (Mechanics of Solids), Vol. 4, 1974, pp. 150-154.

60. Haug, E. J., "Optimization of Distributed Parameter Structures with Repeated Eigenvalues," <u>New Approaches to Nonlinear Problems in Dynamics,</u> (P. J. Holmes, ed.), to appear in 1980.

61. Choi, K. K., Haug, E. J., and Rousselet, B., "Optimization of Structures with Repeated Eigenvalues," 1980, see Reference 286.

62. Rousselet, B., "Singular Dependence of Repeated Eigenvalues," 1980, see Reference 286.

63. Haug, E. J. and Rousselet, B., "Design Sensitivity Analysis in Structural Mechanics I: Static Response Variations," <u>Journal of Structural Mechanics,</u> Vol. 8, No. 1,, 1980, pp. 17-41.

64. Haug, E. J. and Rousselet, B., "Design Sensitivity Analysis in Structural Mechanics II: Eigenvalue Variations," <u>Journal of Structural Mechanics,</u> Vol. 8, 1980, to appear.

65. Thompson, J. M. T. and Lewis, G. M., "On the Optimum Design of Thin-walled Compression Members," <u>Journal of Mechanics and Physics of Solids,</u> Vol. 20, 1972, pp. 101-109.

66. Thompson, J. M. T., "Optimization as a Generator of Structural Instability," <u>International Journal of Mechanical Science,</u> Vol. 14, No. 9, 1972, pp. 627-630.

67. Thompson, J. M. T. and Supple, W. D., "Erosion of Optimum Design by Compound Branching Phenomena," <u>Journal of Mechanics and Physics of Solids,</u> Vol. 21, No. 3, 1972, pp. 135-144.

68. Thompson, J. M. T. and Hunt, G. W., "Dangers of Structural Optimization," <u>Engineering Optimization,</u> Vol. 1, 1974, pp. 99-110.

69. Frauenthal, J. E., "Constrained Optiumal Design of Circular Plates Against Buckling," <u>Journal of Structural Mechanics,</u>, Vol. 1, No. 2, 1972, pp. 159-186.

70. Andreev, L. V., Mossakovsii, V. I., and Obodan, N. I., "On Optimal Thickness of a Cylindrical Shell Loaded by External Presure," <u>PMM,</u> Vol. 36, No. 4, 1972, pp. 677-685.

71. Manevich, A. I. and Kaganov, M. Ye., "Stability and Weight Optimization of Reinforced Spherical Shells Under External Pressure," <u>Prikladnay Mekhanika,</u> Vol. 9, No. 1, 1973, pp. 20-26.

72. Zyczkowski, M. and Kruzelecki, "Optimal Design of Shells with Respect to Their Stability," <u>Optimization in Structural Design,</u> (ed. Sawczuk and Mroz), Springer-Verlag, New York, 1975, pp. 229-247.

73. Tadjbakhsh, I. and Farshad, M., "On Conservatively Loaded Funicular Arches and Their Optimal Design," <u>Optimization in Structural Design,</u> (ed. Sawczuk and Mroz), Springer-Verlag, New York, 1975, pp. 215-228.

74. Niordson, F. I., "On the Optimal Design of a Vibrating Beam," *Quarterly Applied Mechanics*, Vol. 23, 1965, pp. 47-53.

75. Turner, M. J., "Design of Minimum Mass Structures with Specified Natural Frequencies," *AIAA Journal*, Vol. 5, No. 3, 1967, pp. 406-412.

76. Taylor, J. E., "Minimum Mass Bar for Axial Vibration at Specified Natural Frequency," *AIAA Journal*, Vol. 5, No. 10, 1967, pp. 1911-1913.

77. Taylor, J. E., "Optimum Design of a Vibrating Bar with Specified Minimum Cross Section," *AIAA Journal*, Vol. 6, No. 7, 1968, pp. 1379-1381.

78. Sheu, C. Y., "Elastic Minimum-weight Design for Specified Fundamental Frequency," *International Journal Solids and Structures*, Vol. 4, 1968, pp. 953-958.

79. Sippel, D. L. and Warner, W. H., "Minimum-mass Design of Multi-element Structures Under a Frequency Constraint," *AIAA Journal*, Vol. II, No. 4, 1973, pp. 483-489.

80. Cardou, A. and Warner, W. H., "Minimum-mass Design of Sandwich Structures with Frequency and Section Constraints," *Journal of Optimization Theory and Applications*, Vol. 14, No. 6, 1974, pp. 633-647.

81. Miele, A., Mangiavacchi, A., Mohanty, B. P., and Wu, A. K., "Numerical Determination of Minimum Mass Structures with Specified Natural Frequencies," *International Journal of Numerical Methods in Engineering*, Vol. 13, No. 2, 1978, pp. 203-228.

82. Brach, R. M., "On the Extremal Fundamental Frequencies of Vibrating Beams," *International Journal of Solid Structures*, Vol. 4, 1968, pp. 667-674.

83. Brach,, R. M., "On Optimal Design of Vibrating Structures," *Journal of Optimization Theory and Applications*, Vol. 11, 1973, pp. 662-667.

84. Vepa, K., "On the Existence of Solutions to Optimization Problems with Eigenvalue Constraints," *Quarterly Applied Mechanics*, Vol. 31, 1973-1974, pp. 329-341.

85. Brach, R. M., "Optimized Design: Characteristic Vibration Shapes and Resinators," *Journal of the Acoustic Society of America*, Vol. 53, No. 1, 1973, pp. 113-119.

86. McCart, B. R., Haug, E. J., and Streeter, T. D., "Optimal Design of Structures with Constraints on Natural Frequency," *AIAA Journal*, Vol. 8, No. 6, 1970, pp. 1012-1019.

87. Haug, E. J., Arora, J. S., and Matsui, K., "A Steepest-Descent Method for Optimization of Mechanical Systems," *Journal of Optimization Theory and Applications*, Vol. 19, No. 3, 1976, pp. 401-424.

88. Pierson, B. L., "An Optimal Control Approach to Minimum-Weight Vibrating Beam Design," *Journal of Structural Mechanics*, Vol. 5, 1977, pp. 147-178.

89. Kamat, M. P. and Simitses, G. J., "Optimal Beam Frequencies by the Finite Element Displacement Method," *International Journal of Solids and Structures*, Vol. 9, 1973, pp. 414-429.

90. Karihaloo, B. L. and Niordson, F. I., "Optimum Design of Vibrating Beams Under Axial Compression," *Archives of Mechanics*, Vol. 24, 1972, pp. 1029-1037.

91. Karihaloo, B. L. and Niordson, F. I., "Optimum Design of Vibrating Cantilevers," *Journal of Optimization Theory and Applications*, Vol. 11, 1973, pp. 638-654.

92. Seiranyan, A. P., "Optimal Beam Design with Limitations on Natural Vibration Frequency and Buckling Load," (in Russian), *MTT (Mechanics of Solids)*, Vol. 11, No. 1, 1976, pp. 147-152.

93. Warner, W. H. and Vavrick, D. J., "Optimal Design in Axial Motion for Several Frequency Constraints," *Journal of Optimization Theory and Applications*, Vol. 15, No. 1, 1975, pp. 159-166.

94. Olhoff, N., "Optimization of Vibrating Beams with Respect to Higher Order Natural Frequencies," *Journal of Structural Mechanics*, Vol. 4, 1976, pp. 87-122.

95. Olhoff, N., "Maximizing Higher Order Eigenfrequencies of Beams with Constraints on the Design Geometry," *Journal of Structural Mechanics*, Vol. 5, 1977, pp. 107-134.

96. Troitskii, V. A., "Optimization of Elastic Bars in the Presence of Free Vibrations," (in Russian), *MTT (Mechanics of Solids)*, Vol. 11, No. 3, 1976, pp. 145-152.

97. Mroz, A. and Rozvany, G. I. N., "Optimal Design of Structures with Variable Support Conditions," *Journal of Optimization Theory Applications*, Vol. 15, 1975, pp. 85-101.

98. Rozvany, G. I. N., "Analytical Treatment of Some Extended Problems in Structural Optimization, Part I and II," *Journal of Structural Mechanics*, Vol. 3, 1974-1975, pp. 359-402.

99. Olhoff, N. and Niordson, F. I., "Some Problems Concerning Singularities of Optimal Beams and Columns," *ZAMM*, Vol. 59, 1979, pp. T16-T26.

100. Olhoff, N., "Optimization of Transversely Vibrating Beams and Rotating Shafts," 1980,, see Reference 286.

101. Kamat, M. P., "Effect of Shear Deformations and Rotary Inertia on Optimum Beam Frequencies," *International Journal of Numerical Methods in Engineering*, Vol. 9, 1975, pp. 51-62.

102. Weisshaar, T. A., "Optimization of Simple Structures with Higher Mode Frequency Constraints," *AIAA Journal*, Vol. 10, No. 5, 1972, pp. 691-693.

103. Armand, J. -L. and Vitte, W. J., *Foundations of Aeroelastic Optimization and Some Applications to Continuous Systems*, Report No. SUDAAR-390, Department of Aeronautics and Astronautics, Stanford University, 1970.

104. Weisshaar, T. A., *An Application of Control Theory Methods to the Optimization of Structures Having Dynamic or Aeroelastic Constraints*, Report. No. SUDAAR 412, Department of Aeronautics and Astronautics, Stanford University, 1970.

105. Vavrick, D. J. and Warner, W. H., "Minimum Mass Design with Torsional Frequency and Thickness Constraints," *Journal of Structural Mechanics*, Vol. 6, No. 2, 1978, pp. 211-231.

106. Vavrick, D. J. and Warner, W. H., "Duality Among Optimal Design Problems for Torsional Vibration," *Journal of Structural Mechanics*, Vol. 6, No. 2, 1978, pp. 233-246.

107. Maday, C. J., "A Class of Minimum Weight Shafts," *ASME Journal of Engineering for Industry*, Vol. 96, No. 1, 1974, pp. 166-170.

108. DeSilva, B. M. E., "Optimal Vibrational Modes of a Disc," *Journal of Sound and Vibration*, Vol. 21, No. 1, 1972, pp. 19-34.

109. DeSilva, B. M. E., "Optimal Control Concepts in the Design of Turbine Discs and Blades," *Shock and Vibration Digest*, Vol. 7, 1975, pp. 63-76.

110. Grinev, V. B. and Garal, Ya. A., "Optimization of the Parameters of Rotating Rods," *Soviet Applied Mechanics*, Vol. 13, No. 9, 1975, pp. 389-393.

111. Olhoff, N., "Optimal Design of Vibrating Circular Plates," *International Journal of Solids Structures*, Vol. 6, 1970, pp. 139-156.

112. Armand, J. -L., "Minimum-Mass Design of a Plate-Like Structure for Specified Fundamental Frequency," *AIAA Journal*, Vol. 9, No. 9, 1971, pp. 1739-1745.

113. Armand, J. -L., *Applications of the Theory of Optimal Control of Distributed-Parameter Systems to Structural Optimization*, NASA Contractor Report No. NASA CR-2044, NASA, Washington, D.C., 1972.

114. Armand, J. -L., "Numerical Solutions in Optimization of Structural Elements," Paper at First International Conference on Computational Methods in Nonlinear Mechanics, Austin, Texas, 1974.

115. Olhoff, N., "Optimal Design of Vibrating Rectangular Plates," International Journal of Solids Structures, Vol. 10, 1974, pp. 93-109.

116. Olhoff, N., "On Singularities, Local Optima and Formation of Stiffeners in Optimal Design of Plates," Optimization in Structural Design, (Ed. Sawczuk and Mroz), Springer-Verlag, 1975, pp. 82-103.

117. Olhoff, N., "Optimal Design of Solid Elastic Plates," 1980, see Reference 286.

118. Cheng, K. -T. Olhoff, N., An Investigation Concerning Optimal Design of Solid Elastic Plates, DCAMM-Rept. No. 174, The Danish Center for Applied Mathematics and Mechanics, 1980.

119. Olhoff, N., Lurie, K. A., Cherkaev, A. V., and Fedorov, A. V., Sliding Regimes and Anisotrophy in Optimal Design of Vibrating Axisymmetric Plates, DCAMM-Rept., The Danish Center for Applied Mathematics and Mechanics, 1980 (to appear).

120. Lurie, K. A. and Cherkaev, A. V., "Prager Theorem Application to Optimal Design of Thin Plates," MTT (Mechanics of Solids), Vol. 11, 1976, pp. 157-159.

121. Seiranyan, A. P., "A Study of an Extremum in the Optimal Problem of a Vibrating Circular Plate," MTT, Vol. 13, No1. 6, 1978, pp. 99-104.

122. Haug, E. J., "A Gradient Projection Method for Structural Optimization," 1980, (see Reference 286).

123. Haug, E. J., Pan, K. C., and Streeter, T. D., "A Computational Method for Optimal Structural Design I: Piecewise Uniform Structures," International Journal of Numerical Methods in Engineering, Vol. 5, 1972, pp. 171-184.

124. Ainola, L. Ia., "On the Inverse Problem of Natural Vibrations of Elastic Shells," PMM, Vol. 35, No. 2, 1971, pp. 358-364.

125. Barnett, R. L., "Minimum Weight Design of Beams for Deflection," Journal of Engineering Mechanics, ASCE, Vol. 87, EM2, 1961, pp. 75-109.

126. Dixon, L. C. W., "Pontryagin's Maximum Principle Applied to the Profile of a Beam," Journal of Royal Aeronautical Society, Vol. 71, 1967, pp. 513-515.

127. Huang, N. C. and Tang, H. T., "Minimum-Weight Design of Elastic Sandwich Beams with Deflection Constraints," Journal of

Optimization Theory and Applications, Vol. 4, No. 4, 1969, pp. 277-298.

128. Prager, W., "Optimal Thermoelastic Design for Given Deflection," *International Journal of Mechanical Science*, Vol. 12, 1970, pp. 705-709.

129. Chern, J. -M., "Optimal Structural Design for Given Deflection in Presence of Body Forces," *International Journal of Solids and Structures*, Vol. 7, 1971, pp. 363-382.

130. Haug, E. J. and Kirmser, P. G., "Minimum Weight Design of Beams with Inequality Constraints on Stress and Deflection," *Journal of Applied Mechanics*, Vol. 34, 1967, pp. 999-1007.

131. Haug, E. J., *Minimum Weight Design of Beam with Inequality Constraints on Stress and Deflection*, Ph.D. Thesis, Kansas State University, 1966.

132. DeSilva, B. M. E., "Application of Pontryagin's Principle to a Minimum Weight Design Problem," *Journal of Basic Engineering*, Vol. 92, 1970, pp. 245-250.

133. Shield, R. T. and Prager, W., "Optimal Structural Design for Given Deflection," *ZAMP*, Vol. 21, 1970, pp. 513-523.

134. Prager, W., "Optimal Design of Statically Determinate Beams for Given Deflection," *Journal of Mechanical Science*, Vol. 13, 1971, p. 893.

135. Prager, W., "Conditions for Structural Optimality," *Computers and Structures*, Vol. 2, 1972, pp. 833-840.

136. Huang, N. -C., "On Principle of Stationary Mutual Complementary Energy and Its Application to Structural Design," *ZAMP*, Vol. 22, 1971, pp. 608-620.

137. Dafalias, Y. F. and Dupuis, G., "Minimum-Weight Design of Continuous Beams Under Displacement and Stress Constraints," *Journal of Optimization Theory and Applications*, Vol. 9, No. 1, 1972, pp. 137-154.

138. Dupuis, G., "Optimal Design of Statically Determinate Beams Subject to Displacement and Stress Constraints," *AIAA Journal*, Vol. 9, No. 5, 1971.

139. Dupuis, G., "An Iterative Approach to Structural Optimization," *International Journal for Numerical Methods in Engineering*, Vol. 4, 1972, pp. 331-336.

140. Bhargava, S. and Duffin, R. J., "Dual Extremum Principles Relating to Optimum Beam Design," *Archives of Rational Mechanics and Analysis*, Vol. 50, 1973, pp. 314-330.

141. Simitses, G. J. and Kotras, T., "The Optimal Euler-Bernoulli Cantilever," ASCE Journal of Engineering Mechanics Division, ASCE, Vol. 101, No. EM6, 1975, pp. 922-929.

142. Distefano, N. and Todeschini, R., "Invariant Imbedding Optimum Beam Design with Displacement Constraints," International Journal of Solids and Structures, Vol. 8, No. 8, 1972, pp. 1073-1088.

143. Huang, N. -C., "Optimal Design of Elastic Beams for Minimum-Maximum Deflection," Journal of Applied Mechanics, Vol. 38, 1971, pp. 1078-1081.

144. Komkov, V. and Coleman, N. P., "An Analytic Approach to Some Problems of Optimal Design of Beams and Plates," Archives of Mechanics, Vol. 27, No. 4, 1975, pp. 565-575.

145. Cinquini, C., "Optimal Elastic Design for Prescribed Maximum Deflection," Journal of Structural Mechanics, Vol. 7, No. 1, 1979, pp. 21-34.

146. Armand, J. -L., "Applications of Optimal Control Theory to Structural Optimization: Analytical and Numerical Approach," Proceedings IUTAM Symposium on Optimization in Structural Design, (Ed. A. Sawczuk and Z. Mroz), Springer-Verlag, 1975, pp. 15-39.

147. Banichuk, N. V., "Game Problems in the Theory of Optimal Design," Optimization in Structural Design, (Ed. Sawozuk and Mroz), Springer-Verlag, New York, 1975, pp. 111-121.

148. Hegemier, G. A. and Tang, H. T., "A Variational Principle, The Finite Element Methods, and Optimal Structural Design for Given Deflection," Optimization in Structural Design, (Ed. Sawczuk and Mroz), Springer-Verlag, New York, 1975, pp. 464-483.

149. Armand, J. -L. and Lodier, B., "Optimal Design of Bending Elements," International Journal of Numerical Methods in Engineering, Vol. 13, 1978, pp. 373-384.

150. Banichuk, N. V., "Optimal Elastic Plate Shapes in Bending Problems," MTT, Vol. 10, No. 5, 1975, pp. 151-158.

151. Banichuk, N. V., Karelishvili, V. M., and Mironov, A. A., "Numerical Solution of Two-Dimensional Optimization Problems for Elastic Plates," MTT, Vol. 12, No. 1, 1977, pp. 65-74.

152. Banichuk, N. V., Karelishvili, V. M., and Mironov, A. A., "Optimization Problems with Local Performance Criteria in the Theory of Plate Bending," MTT, Vol. 13, No. 1, 1978, pp. 116-122.

153. Banichuk, N. V., "Design of Plate for Minimum Deflection and Stress," 1980, see Reference 286.

154. Simitses, G. J., "Optimal versus The Stiffened Circular Plate," AIAA Journal, Vol. 11, No. 10, 1973, pp. 1409-1412.

155. Sheu, C. Y. and Prager, W., "Minimum-Weight Design with Piecewise Constant Specific Stiffness," *Journal of Optimization Theory and Applications*, Vol. 2, 1968, p. 179-189.

156. Prager, W., "Optimality Criteria in Structural Design," *Proceedings of National Academy of Sciences*, Vol. 61, 1968, p. 794-797.

157. Hegemier, G. A. and Prager, W., "On Michell Trusses," *International Journal of Mechanical Sciences*, Vol. 11, 1969, p. 209.

158. Martin, J. B., "The Optimal Design of Beams and Frames with Compliance," *International Journal of Solids and Structures*, Vol. 7, 1971, pp. 63-81.

159. Huang, N.-C., "Optimal Design of Elastic Structures for Maximum Stiffness," *International Journal of Solids and Structures*, Vol. 4, 1968, pp. 689-700.

160. Haung, N.-C. and Sheu, C. Y., "Optimal Design of Elastic Circular Sandwich Beams for Minimum Compliance," *Journal of Applied Mechanics*, Vol. 37, 1970, p. 569.

161. Chern, J. M. and Prager, W., "Optimal Design for Precribed Compliance Under Alternative Loads," *Journal of Optimization Theory and Applications*, Vol. 5, 1970, pp. 424-431.

161a. Chern, J. M., "Optimal Design of Beams for Alternative Loads and Constraints on Generalized Compliance and Stiffness," *International Journal of Mechanical Science*, Vol. 13, No. 8, 1971, pp. 661-674.

161b. Prager, W., "Conditions for Structural Optimality," *Computers and Structures*, Vol. 1, 1972, pp. 833-840.

162. Save, M., "A General Criterion for Optimal Structural Design," *Journal of Optiization Theory and Applications*, Vol. 15, 1975, pp. 119-129.

163. Masur, E. F., "Optimum Stiffness and Strength of Elastic Structures," *ASCE Journal of Engineering Mechanics Division*, Vol. 95, No. EM5, 1970, pp. 621-640.

163a. Masur, E. F., "Optimal Structural Design for a Discrete Set of Available Structural Members," *Computer Methods in Applied Mechanics and Engineering*, Vol. 3, 1976, pp. 195-207.

163.b. Masur, E. F., "Optimality in the Presence of Discreteness and Discontinuity," *Optimization in Structural Design*, (Ed. Sawczuk and Mroz), Springer-Verlag, 1975, pp. 441-453.

164. Mroz, Z., "Multiparameter Optimal Design of Plates and Shells," *Journal of Structural Mechanics*, Vol. 1, No. 3, 1973, pp. 371-392.

165. Mroz, Z. and Rozvany, G. I. N., "Optimal Design of Structures with Variable Support Conditions," Journal of Optimization Theory and Applications, Vol. 15, No. 1, 1975, pp. 85-101.

166. Reiss, R., "Optimal Compliance Criterion for Axisymmetric Solid Plates," International Journal of Solids Structures, Vol. 12, 1976, pp. 319-329.

167. Chern, J. -M. and Prager, W., "Optimal Design of Rotating Disk for Given Radial Displacement of Edge," Journal of Optimization Theory and Applications, Vol. 6, No. 2, 1970, pp. 161-170.

168. Chern, J. -M., "Optimal Thermo-Elastic Design for Given Deformation," Journal of Applied Mechanics, Vol. 38, No. 2, 1971, pp. 538-540.

169. Fuchs, M. B. and Brull, M. A., "A New Strain Energy Theorem and Its Use in the Optimum Design of Continuous Beams," Computers and Structures, Vol. 10, 1979, pp. 647-657.

170. Icerman, L. J., "Optimal Structural Design for Given Dynamic Deflection," International Journal of Solids and Structures, Vol. 5, 1969, pp. 473-490.

171. Mroz, Z., "Optimal Design of Structures Subjected to Dynamic, Harmonically-Varying Loads," ZAMM, Vol. 50, 1970, pp. 303-309.

172. Plaut, R. H., "Optimal Structural Design for Given Deflection Under Periodic Loading," Quarterly of Applied Mathematics, Vol. 29, 1971, pp. 315-318.

173. Plaut, R. H., "Approximate Solutions to Some Static and Dynamic Optimal Structural Design Problems," Quarterly of Applied Mathematics, Vol. 31, 1973, pp. 535-539.

174. Huang, N. C., "Minimum-Weight Design of Vibrating Elastic Structures with Dynamic Deflection Constraint," Journal of Applied Mechanics, Vol. 43, 1976, pp. 171-180.

175. Rockenback, P. C., Minimum-Mass Response-Constrained Design of Vibrating Sandwich Beams, Report R-604, Coordinated Science Laboratory, University of Illinois, Urbana, 1973.

176. McNamara, R. J., "Turned Mass Dampers for Buildings," Journal of Structural Division, ASCE, Vol. 103, No. ST9, 1977, pp. 1785-1798.

177. Petersen, N. R., "Design of Large Scale Turned Mass Dampers," Preprint 3578, ASCE National Meeting, Boston, April 1979.

178. Wiesner, K. B., "Turned Mass Dampers to Reduce Building Wind Motion," Preprint 3510, ASCE National Meeting, Boston, April 1979.

179. Rao, S. S., "Structural Optimization Under Shock and Vibration Environment," The Shock and Vibration Digest, Vol. 11, No. 2, 1979, pp. 3-12.

180. Magne, R. W., "Optimization Techniques for Shock and Vibration Isolator Development," Shock and Vibration Digest, Vol. 11, No. 10, 1979, pp. 25-33.

181. Brach, R. M., "Minimum Dynamic Response for a Class of Simply Supported Beam Shapes," International Journal Mechanical Science, Vol. 10, 1968, pp. 429-439.

182. Brach, R. M., "Optimum Design of Beams for Sudden Loading," Journal of Engineering Mechanics Division, ASCE, Vol. 96, No. EM6, 1968, pp. 1395-1407.

183. Plaut, R. H., "On Minimizing the Response of Structures to Dynamic Loading," ZAMP, Vol. 21, 1970, pp. 1004-1010.

184. Pochtman, Y. M., "Optimization of Structures with Constraints on Dynamic and Frequency Characteristics," Doklady Akademii Nauk SSSR, (in Russian), Vol. 203, No. 2, 1972, pp. 307-308; English Translation NASA TT F-14, NASA, 1972.

185. Fox, R. L. and Kapoor, M. P., "Structural Optimization in the Dynamics Response Regime: A Computational Approach," AIAA Journal, Vol. 8, No. 10, 1970, pp. 1798-1804.

186. Cassis, J. H. and Schmit, L. A., "Optimum Design with Dynamic Constraints," Journal of Structural Division, ASCE, Vol. 102, No. ST10, 1976, pp. 2053-2071.

187. Feng, T. -T., Arora, J. S., and Haug, E. J., "Optimal Structural Design Under Dynamic Loads," International Journal of Numerical Methods in Engineering, Vol. 11, 1977, pp. 39-52.

188. Arora, J. S. and Haug, E. J., "Optimum Structural Design with Dynamic Constraints," ASCE Journal of Structural Division, Vol. 103, No. ST10, 1977, pp. 2071-2074.

189. Haug, E. J. and Feng, T. -T., "Optimization of Distributed Parameter Structures Under Dynamic Loads," Control and Dynamic Systems, (Ed. C. T. Leandes), Vol. 13, 1977, pp. 207-246.

190. Haug, E. J. and Feng, T. -T., "Optimal Design of Dynamically Loaded Continuous Structures," International Journal of Numerical Methods in Engineering, Vol. 12, 1978, pp. 299-307.

191. Haug, E. J., Arora, J. S., and Feng, T. -T., "Sensitivity Analysis and Optimization of Structures for Dynamic Response, Journal of Mechanical Design, Vol. 100, 1978, pp. 311-318.

192. Hsiao, M. H., Haug, E. J., and Arora, J. S., "A State of Space Method for Optimal Design of Vibration Isolators," Journal of Mechanical Design, Vol. 101, 1979, pp. 309-314.

193. Haug, E. J. and Arora, J. S., "Distributed Parameter Structural Optimization for Dynamic Response," 1980, see Reference 286.

194. Kato, B., Nakrmara, Y., and Anraku, H., "Optimum Earthquake Design of Shear Buildings," Journal of Engineering Mechanics Division, ASCE, Vol. 98, No. EM4, 1972, pp. 892-909.

195. Venkayya, V. B. and Khot, N. S., "Design of Optimum Structures to Impulse Type Loading," AIAA Journal, Vol. 13, 1975, pp. 989-994.

196. Cheng, F. Y. and Botkin, M. E., "Nonlinear Optimum Design of Dynamic Damped Frames," Journal of Structural Division, ASCE, Vol. 102, No. ST3, 1976, pp. 609-628.

197. Cheng, F. Y. and Srifuengfung, D., "Earthquake Structural Design Based on Optimality Criterion," Sixth World Conference on Earthquake Engineering, Vol. 5, Earthquake Resistant Design, 1977.

198. Cheng, F. Y. and Srifuengfung, D., "Optimal Structural Design for Simultaneous Multicomponent Static and Dynamic Input," International Journal of Numerical Methods in Engineering, Vol. 13, 1978, pp. 353-371.

199. Ray, D., Pister, K. S., and Chopra, A. K., "Optimum Design of Earthquake-Resistant Shear Buildings," EERC 74-3, Earthquake Engineering Research Center, U.S. Berkeley, January 1974.

200. Vitiello, E. and Pister, K. S., Applications of Reliability-Based Global Cost Optimization to Design of Earthquake-Resistant Structures, Report No. ERC 74-10, University of California, Berkeley, August 1974.

201. Walker, N. D. and Pister, K. S., Study of a Method of Feasible Directions for Optimal Elastic Design of Framed Structures Subjected to Earthquake Loading, EERC 75-39, Earthquake Engineering Research Center, U.C. Berkeley, December 1975.

202. Ray, D., Sensitivity Analysis for Hysteretic Dynamic Systems: Application to Earthquake Engineering, EERC 74-5, Earthquake Engineering Research Center, U.S. Berkeley, April 1974.

203. Ray, D., Pister, K. S., and Polak, E., "Sensitivity Analysis for Hysteretic Dynamic Systems: Theory and Applications," Computer Methods in Applied Mechanics and Engineering, Vol. 14, 1978, pp. 179-208.

204. Bhatti, M. A., Pister, K. S., and Polak, E., Optimal Design of an Earthquake Isolation System, Report No. EERC 78-22, University of California, Berkeley, October 1978.

205. Bhatti, M. A., Polak, E., and Pister, K. S., OPTDYN--A General Purpose Optimization Program for Problems With or Without Dynamic Constraints, Report No. EERC 79-16, University of California, Berkeley, July 1979.

206. Bhatti, M. A., Optimal Design of Localized Nonlinear Systems with Dual Performance Criteria Under Earthquake Excitations, Report No. EERC 79-15, University of California, Berkeley, July 1979.

207. Pister, K. S., "Optimal Design of Structures Under Dynamic Loading," 1980, see Reference 286.

208. Polak, E., "Algorithms for Optimal Design," 1980, see Reference 286.

209. Bhatti, M. A., Essebo, T., Nye, W., Pister, K. S., Polak, E., Sangiovanni-Vincentelli, and Titz, A., "A Software System for Optimization-Based Interactive Computer-Aided Design," 1980, see Reference 286.

210. Bhatti, M. A. and Pister, K. S., "Applications of Optimal Design to Structures Subjected to Earthquake Loading, 1980, see Reference 286.

211. Levy, H. J. and Wolf, B. M., "Fully Stressed Dynamically Loaded Structures," ASME Paper 74-WA/DE-19, ASME, New York, 1974.

212. Komkov, V., Optimal Control Theory for the Damping of Elastic Vibrations of Simple Elastic Systems, Springer-Verlag, Berlin, 1972.

213. Komkov, V. and Coleman, N., "Optimality of Design and Sensitivity Analysis of Beam Theory," International Journal of Control, Vol. 18, No. 4, 1973, pp. 731-740.

214. Therman, K., "Optimal Design Criteria of Dynamically Loaded Elastic Structures," Optimization in Structural Design, (Ed. Sawczuk and Mroz), Springer-Verlag, New York, 1975, pp. 152-167.

215. Zienkiewicz, O. C. and Campbell, J. S., "Shape Optimization and Sequential Linear Programming," Optimum Structural Design, (Ed. Gallagher, R. H. and Zienkiewica, O. C.), Wiley, New York, 1973, pp. 109-126.

216. Ramakrishnan, C. V. and Francavilla, A., "Structural Shape Optimization Using Penalty Functions," Journal of Structural Mechanics, Vol. 3, No. 4, 1975, pp. 403-432.

217. Francavilla, A., Ramakrishnan, C. V., and Zienkiewicz, "Optimization of Shape to Minimize Stress Concentration," Journal of Strain Analysis, Vol. 10, 1975, pp. 63-70.

218. Tvergaard, V., "On the Optimum Shape of a Fillet in a Flat Bar with Restrictions," Optimization in Structural Design, (Ed. Sawczuk and Mroz), Springer-Verlag, New York, 1975, pp. 181-195.

219. Kristensen, E. S. and Madsen, N. F., "On the Optimum Shape of Fillets in Plates Subjected to Multiple In-Plane Loading Cases," International Journal of Numerical Methods in Engineering, Vol. 10, 1976, pp. 1007-1019.

220. Bhavikatti, S. S. and Ramakrishnan, C. V., "Optimum Design of Fillets in Flat and Round Tension Bars," ASME Paper, 77-DET-45, 1977.

221. Chun, Y. W. and Haug E. J., "Two Dimensional Shape Optimal Design," International Journal of Numerical Methods in Engineering, Vol. 13, 1978, pp. 311-336.

222. Chun, Y. W. and Haug, E. J., "Shape Optimal Design of an Elastic Body of Revolution," Preprint No. 3516, ASCE Annual Meeting, Boston, April, 1979.

223. Rousselet, B. and Haug, E. J., "Design Sensitivity Analysis in Structural Mechanics III: Shape Variation," 1980, see Reference 286.

224. Henry, A. S., The Analytic Design of Torsion Members, Ph.D. Thesis, University of Iowa, 1971.

225. Banichuk, N. V., "Optimization of Elastic Bars in Torsion," International Journal of Solids and Structures, Vol. 12, 1976, pp. 275-286.

226. Banichuk, N. V., "On a Variational Problem with Unknown Boundaries and the Determination of Optimal Shapes of Elastic Bodies," PMM, Vol. 39, No. 6, 1975, pp. 1037-1047.

227. Kurshin, L. M. and Onoprienko, P. N., "Determination of the Shapes of Doubly-Connected Bar Sections of Maximum Torsional Stiffness," PMM, Vol. 40, No. 6, 1976, pp. 1020-1026.

228. Banichuk, N. V., "On a Two Dimensional Optimization Problem in Elastic Bar Torsion Theory," Soviet Applied Mechanics, Vol. 11, No. 5, 1976, pp. 38-44.

229. Gurvitch, E. L., "On Isoparimetric Problems for Domains with Partly Known Boundaries," Journal of Optimization Theory and Applications, Vol. 20, No. 1, 1976, pp. 65-79.

230. Neuber, H., "Der Zugbeanspruchte Flachstab Mit Optimalem Querschnittubergang," Forsch Ingenieurwesen, Vol. 35, 1966, pp. 29-30.

231. Neuber, H., "Zur Optimierun Der Spannungskonzentration," in Continuum Mechanics and Related Problems of Analysis, Nauka, Moscow, 1972, pp. 375-380.

232. Cherepanov, G. P., "Inverse Problems of the Plane Theory of Elasticity," PMM, Vol. 38, No. 6, 1974, pp. 963-979.

233. Wheeler, L., "On the Role of Constant-Stress Surfaces in the Problem of Minimizing Elastic Stress Concentration," International Journal of Solids and Structures, Vol. 12, 1976, pp. 779-789.

234. Banichuk, N. V., "Optimality Conditions in the Problem of Seeking the Hole Shapes in Elastic Bodies," PMM, Vol. 41, No. 5, 1977, pp. 920-925.

235. Banichuk, N. V. "Optimizing Hole Shape in Plates Working in Bending," Soviet Applied Mechanics, Vol. 12, No. 3, 1977, pp. 72-78.

236. Banichuk, N. V. and Karihaloo, B. L., "Minimum-Weight Design of Multi-Purpose Cylindrical Bars," International Journal of Solids and Structures, Vol. 12, 1976, pp. 267-273.

237. Parbery, R. D. and Karihaloo, B. L., "Minimum-Weight Design of Hollow Cylinders for Given Lower Bounds on Torsional and Flexural Rigidities," International Journal of Solids and Structures, No. 13, 1977, pp. 1271-1280.

238. Cherkaev, A. V., "On the Question of Formulating the Problem of Optimal Design of Freely Oscillating Structures," PMM, Vol. 42, No. 1, 1978, pp. 194-197.

239. Durelli, A. J. and Rajaiah, K., "Optimum Hole Shapes in Finite Plates Under Uniaxial Load," Journal of Applied Mechanics, Vol. 46, 1979, pp. 691-695.

240. Cea, J., "Solution of a Model Problem by Variational Methods and Examples of Problems of Shape Optimal Design," 1980, see Reference 286.

241. Cea, J., "Definition of Boundaries for Shape Design," 1980, see Reference 286.

242. Cea, J., "Continuous Steepest Descent in Hilbert Space and 'Domain Spaces'," 1980, see Reference 286.

243. Zolesio, J. P., "The Material Derivative (Or Speed) Method for Shape Optimization," 1980, see Reference 286.

244. Zolesio, J. P., "Speed Method in Several Examples," 1980, see Reference 286.

245. Rousselet, B., "Implementation of Shape Optimal Design Algorithms," 1980, see Reference 286.

246. Cea, J., "Other Methods in Shape Optimal Design," 1980, see Reference 286.

247. Rousselet, B., "Dependence of Eigenvalues on Shape," 1980, see Reference 286.

248. Prager, W. and Shield, R. T., "Optimal Design of Multi-Purpose Structures," International Journal of Solids and Structures, Vol. 4, 1968, pp. 469-475.

249. Martin, J. B., "Optimal Design of Structures for Multipurpose Loading," Journal of Optimization Theory and Applications, Vol. 6, No. 1, 1970, pp. 22-40.

250. Chern, J. -M. and Martin, J. B., "The Multipurpose Optimal Design of Elastic Structure with a Piecewise Uniform Cross-Section," ZAMP, Vol. 22, No. 5, 1971, pp. 834-855.

251. Karihaloo, B. L. and Niordson, F. I., "Optimum Design of Vibrating Beams Under Axial Compression," Archives of Mechanics, Vol. 24, 1972.

252. Sherman, Z. and Wang, P. -C., "Volume Minimization of Thin Plates Subject to Constraints," ASCE Journal of Engineering Mechanics Division, Vol. 97, No. EM3, 1971, pp. 741-754.

253. Sherman, Z., "Weight Minimization Axisymmetric Clamped Plates Subject to Constraints," International Journal of Solids and Structures, Vol. 9, 1973, pp. 279-290.

254. Karihaloo, B. L., "Optimal Design of Multi-Purpose Tie-Beams," Journal of Optimization Theory and Applications, Vol. 27, No. 3, 1979, pp. 427-438.

255. Karihaloo, B. L. and Parbery, R. D., "Optimal Design of Multi-Purpose Beam-Column," Journal of Optimization Theory and Applications, Vol. 27, No. 3, 439-448.

256. Karihaloo, B. L., "Optimal Design of Multi-Purpose Structures", Journal of Optimization Theory and Aplications, Vol. 27, No. 3, pp. 449-461.

257. Karihaloo, B. L. and Wood, G. L., "Optimal Design of Multipurpose Sandwich Tie-Columns," ASCE Journal of Engineering Mechanics Division, Vol. 105, No. EM3, 1979, pp. 465-469.

258. Haug, E. J. and Komkov, V., "Sensitivity Analysis in Distributed-Parameter Mechanical System Optimization," Journal of Optimization Theory and Applications, Vol. 23, No. 3, 1977, pp. 445-464.

259. Haug, E. J. and Arora, J. S., "Design Sensitivity Analysis of Elastic Mechanical Systems," Computer Methods in Applied Mechanics and Engineering, Vol. 15, 1978, pp. 35-62.

260. Huang, N. C., "Effect of Shear Deformation on Optimal Design of Elastic Beams," International Journal of Solid and Structures, Vol. 7, 1971, pp. 321-326.

261. Seiranyan, A. P., "Elastic Plates and Beams of Minimum Weight with Several Types of Bending Loads," MTT, Vol. 8, No. 5, 1973, pp. 83-89.

262. Seiranyan, A. P., "Optimal Beam Design With Limitations on Natural Vibration Frequency and Buckling Load," MTT, Vol. 11, No. 1, 1976, pp. 133-138.

263. Seiranyan, A. P., "Quasioptimal Solutions of Optimal Design Problems with Various Constraints," *Soviet Applied Mechanics*, Vol. 13, No. 6, 1977, pp. 544-550.

264. Gura,, N. M. and Seiranyan, A. P., "Optimum Circular Plate with Cosntraints on the Rigidity and Frequency of Natural Oscillations," *MTT*, Vol. 12, No. 1, 1977, pp. 129-136.

265. Seiranyan, A. P., "Homogeneous Functionals and Structural Optimization Problems," *International Journal of Solids and Structures*, Vol. 15, 1979, pp. 749-759.

266. Velte, W. and Villaggio, P., "Are the Optimum Problems in Structural Design Well Posed?," *Archives of Rational Mechanics and Analysis*, to appear in 1980.

267. Klosowicz, B. and Lurie, K. A., "On the Optimal Nonhomogeneity of a Torsional Elastic Bar," *Archives of Mechanics*, Vol. 24, No. 2, 1971.

268. Banichuk, N. V., "One Extremum Problem for a System with Distributed Parameters and Determination of the Optimal Properties of an Elastic Medium," DALK, AM SSSR, Vol. 242, No. 5, 1978.

269. Banichuk, N. V., "Optimal Anisotropy of Rods in Torsion," *MTT*, No. 4, 1978.

270. Banichuk, N. V., "Optimization of Anisotropic Properties of Deformable Media in Plane Problems of Elasticity," *MTT*, Vol. 14, No. 1, 1979, pp. 63-68.

271. Hirano, Y., "Optimum Design of Laminated Plates Under Axial Compression," *AIAA Journal*, Vol. 17, No. 9, 1979, pp. 1017-1019.

272. Banichuk, N. V., "Minimax Approach to Structural Optimization Problems," *Journal of Optimization Theory and Applications*, Vol. 20, No. 1, 1976, pp. 111-118.

273. Banichuk, N. V., "On A Game Problem of Optimizing Elastic Bodies," *Soviet Math Dakl*, Vol. 226, No. 3, 1976, pp. 117-120.

274. Aristov, M. V. and Troitskii, V. A., "Elastic Circular Plate of Minimum Weight," *MTT*, No. 3, 1975, pp. 153-156.

275. Samsonov, A. M., "Optimum Location of Thin Elastic Rib on Elastic Plate," *MTT*, Vol. 13, No. 1, 1978, pp. 121-129.

276. Soldovnikov, V. M., "Optimization of Elastic Shells of Revolution," *PMM*, Vol. 42, No. 3, 1978, pp. 535-544.

277. Olhoff, N. and Taylor, J. E., "On Optimal Structural Remodeling," *Journal of Optimization Theory and Applications*, Vol. 27, No. 4, 1979, pp. 571-581.

278. Taylor, J. E., "Optimal Remodeling Theory and Applications," 1980, see Reference 286.

279. Batterman, S. C. and Pavicic, Nick, "Optimum Design of Fiber Reinforced Shells of Revolution," *Optimization in Structural Design*, Springer-Verlag, New York, 1975.

280. Taylor, J. E., "On Variational Formulations for Structures Design Problems," *Optimization in Structural Design*, (Ed. Sawczuk and Mroz), Springer-Verlag, New York, 1975, pp. 60-67.

281. Rozvany, G. I. N., "Analytical Treatment of Some Extended Problems in Structural Optimization," *Journal of Structural Mechanics*, Vol. 3, No. 4, 1974-1975, pp. 359-385.

282. Conry, T. F. and Seireg, A., "A Mathematical Programming Method for Design of Elastic Bodies in Contact," *Journal of Applied Mechanics*, Vol. 48, 1971, pp. 387-392.

283. Haug, E. J. and Kwak, B. M., "Contact Stress Minimization by Contour Design," *International Journal of Numerical Methods in Engineeering*, Vol. 12, 1978, pp. 917-930.

284. Lukasiewicz, S. A., "Optimum Design in Junction and Contact Problems," *Colloquium No. 110, Contact Problems and Load Transfer in Mechanical Assemblages*, Linkoping, Sweden, 1978.

285. Benedict, R. L., "Optimal Design for Elastic Bodies in Contact," 1980, see Reference 286.

286. *Proceedings NATO Advanced Study Institute on Distributed Parameter Optimization*, May-June 1980, Iowa City, two volumes to be published by Sijthoff and Nordhoff.

APPENDICES

APPENDIX A

ANALYTICAL METHODS IN STRUCTURAL OPTIMIZATION: AN UPDATE[1]

Plastic Design

Cinquini, et al., (1) discussed recently the optimal design of circular plates and Cinquini and Sachi general problems in the optimal elastic and plastic design of structures (2).

Optimal Layout Problems

The optimal layout of trusses was investigated further by Prager, who proposed a geometrical method (3) and also developed an ingenious technique for minimum weight trusses whose design is restricted to a specified number of members (4, 5).

Layout optimization of grillages can not be solved analytically in a closed form for all combinations of boundary conditions (6) since the problems of free edges has been resolved (7). Additional recent developments have been the inclusion of the shear force in the specific cost function (8) and automatic generation and plotting of optimal beam layouts by purely analytical methods on the computer (9).

The curved equivalent of grillages, **viz.**, optimal shell grids were considered recently by Prager and Rozvany (10).

Allowance for Selfweight

Selfweight becomes an important factor when the span of the structure is long. A simple modification of the Prager-Shield optimality criterion for inclusion of selfweight was proposed by Rozvany (11). Optimization of arches (12) and spherical cupolas (13) for selfweight and weight of roofsheeting was considered recently by Hill, et al., and Prager and Rozvany. More recently, the role of selfweight (14) and alternate loads (15) in shell grid optimization were explored.

Superposition Principles

A very powerful superposition principle for two alternate loads was developed quite independently by Hemp (16), by Nagtegaal and Prager (17, 18), and by Spillers and Lev (19-22). This principle was extended recently to optimal design considering an arbitrary number of alternate loads and applied to Michell trusses and grillages (23).

---

[1]The reader is referred to a review published in Applied Mechanics Reviews (AMR), Vol. 30, No. 11, November 1977, in which Rozvany and Mroz provides an excellent coverage of analytical methods. The review does not emphasize optimization of distributed parameter structures, as does Chapter 9. The AMR review and Chapter 9 together may be considered to cover analytical methods very comprehensively. This appendix is provided as a supplement, or a short update to the AMR review, and was contributed by G. I. N. Rozvany.

REFERENCES

1. Cinquini, C., Lamblin, D., and Guerlement, G., "Variational Formulation of the Optimal Plastic Design of Circular Plates," Computer Methods of Applied Mechanical Engineering, 11, 1, April 1977, pp. 19-30.

2. Cinquini, C. and Sachi, G., "Problems of Optimal Design for Elastic and Plastic Structures," Rep. 48, Inst. Sci. Techn. Univ. Pavia.

3. Prager, W., "Geometric Discussion of the Optimal Design of a Simple Truss," J. Struct. Mech. 4, 1, 1976, pp. 57-64.

4. Prager, W., "Optimal Layout of Cantilever Trusses," JOTA, 23, 1, September 1977, pp. 111-117.

5. Prager, W., "Nearly Optimal Design of Trusses," Computers and Structures 8, 1978, pp. 451-454.

6. Prager, W. and Rozvany, G. I. N., "Optimal Layout of Grillages," J. Struct. Mech. 5, 1977, pp. 1-18.

7. Hill, R. and Rozvany, G. I. N., "Optimal Beam Layouts: The Free Edge Paradox," J. Appl. Mech. ASME 44, 4, December 1977, pp. 696-700.

8. Rozvany, G. I. N., "Optimal Beam Layouts: Allowance for Cost Shear," Computer Methods of Applied Mechanical Engineering, 19, 1, 1979, pp. 49-58.

9. Rozvany, G. I. N. and Hill, R., "A Computer Algorithm for Deriving Analytically and Plotting Optimal Structural Layout," Proc. NASA Symp., Washington, 1978, 295-300 and Comp. Struct. 10, 1-2, April 1979, pp. 295-300.

10. Rozvany, G. I. N. and Prager, W., "A New Class of Optimization Problems: Optimal Archgrids," Computer Methods of Applied Mechanical Engineering, 19, 1, 1979, pp. 127-150.

11. Rozvany, G. I. N., "Optimal Plastic Design: Allowance for Self-weight," J. Eng. Mech. Div. ASCE, 103, EM6, December 1977, pp. 1165-1170.

12. Hill, R., Rozvany, G. I. N., Wang, C. M., and Leong, K. H., "Optimization, Spanning Capacity and Cost Sensitivity of Fully Stressed Arches," J. Struct. Mech., (in press).

13. Prager, W. and Rozvany, G. I. N., "Optimal Spherical Cupola of Uniform Strength," Ingenieur-Archiv, (in press).

14. Rozvany, G. I. N., Wang, C. M., and Dow, M., "Arch Optimization Using Prager-Shield Criteria," Journal of Engineering Mechanics Division, ASCE (in press).

15. Rozvany, G. I. N., Nakamura, H., and Kuhnell, B. T., "Optimal Archgrids: Allowance for Selfweight," Computer Methods of Applied Mechanical Engineering, (in press).

16. Hemp, N. S., Optimum Structures, Clarendon Press, Oxford, 1973.

17. Nagtegaal, J. C. and Prager, W., "Optimal Layout of a Truss for Alternative Loads," International Journal of Mechanical Sciences, Vol. 15, pp. 583-592, 1973.

18. Nagtegaal, J. C., "A Superposition Principle in Optimal Plastic Design for Alternative Loads," International Journal of Solids Structures, Vol. 9, pp. 1465-1471, 1973.

19. Spillers, W. R. and Lev., O. E., "Design for Two Loading Conditions," International Journal of Solids Structures, Vol. 7, pp. 1261-1267, 1971.

20. Spillers, W. R., "A Note on the Decommposition of an Absolute Value Linear Programming Problem," Quarterly of Applied Mathematics, Vol. 29, No. 4, January 1972, pp. 541-544.

21. Lev, O. E., "A Structural Optimization Solution to a Branch-and-Bound Problem," Quarterly of Applied Methematics, Vol. 34, pp. 365-371, 1977.

22. Rozvany, G. I. N. and Hill, R. D., "Optimal Plastic Design: Superposition Principles and Bounds on the Minimum Cost," Computer Methods in Applied Mechanical Engineering.

APPENDIX B

OPTIMALITY CRITERION APPROACH: RECENT ADVANCES

N. S. Khot* and V. B. Venkayya*

B.1 INTRODUCTION

This appendix briefly discusses some recent developments in the algorithms based on the optimality criterion approach to design a minimum weight structure. In the optimality criterion approach the optimality conditions that the minimum weight design has to satisfy are first derived, and then an iterative algorithm is developed in order to obtain a design satisfying these conditions. The nature of the optimality criterion depends on the type of constraints imposed on the structure. The constraints may include a maximum allowable stress, a limitation on displacements, dynamic stiffness, system stability, local element buckling, maximum and minimum size limitations.

In the mathematical form the optimality criterion is equivalent to the Kuhn-Tucker conditions of nonlinear programming. However, for a particular type of constraints, it gives information on the distribution of the energy in the structure necessary to have a minimum weight design. The nature of the energy depends on the type of constraint imposed on the structure. For example in the case of displacement and stress constraints, the energy is the virtual strain energy due to the virtual load system. When the structure is designed to satisfy system stability, the energy is the strain energy in the buckled mode, and so on.

The optimality criterion approach can be considered an "indirect method," since the objective is to obtain a design satisfying the optimality conditions and by so doing indirectly minimize the weight of the structure. In this work, discussion of the optimality criterion method is restricted to built up structures which are idealized by discrete elements and analyzed by the finite element method. A discussion on the optimality criterion method as applied to continuum structures may be found in References 1 and 2.

In a structural optimization problem it is generally assumed that the geometry and the loads applied to the structures are known, and the design variables are the member sizes. There are two main steps involved in structural optimization. These are 1) to analyze the structure in order to find the response of the structure to the applied loads and 2) to redistribute the material so that the weight of the structure is reduced and the constraints are satisfied after each iteration. In the discretized structure the analysis is performed bythe finite element method. Redistribution of the material in the optimality criterion approach is achieved by using the recurrence

---

*Air Force Flight Dynamics Laboratory, AFFDL/FBR, Wright-Patterson AFB, Ohio

relation, obtained from the optimality criterion, to modify the design variables. The optimality criterion is derived by differentiating the Lagrangian with respect to the design variables. Since the structures are generally indeterminate and the constraints nonlinear, the structural optimization algorithms are iterative in nature.

The basic idea of the optimality criterion methods for different types of constraints is discussed in References 3-5. In Reference 6 the optimality criterion and the recurrence relations are derived by a different approach. Application of the method to stress and displacement constraint problems is presented in References 7-23. A system stability as the design constraint is considered in References 24-25. Design of a structure with a dynamic load system is discussed in References 26-31. Computer programs based on the optimality criterion approach are documented in References 32-35. These are not all the references in this field, additional references may be found in Reference 36.

In this appendix, a short summary is presented of the optimality criteria for various design constraints, but only the algorithms for designing a structure with stress and displacement constraints are discussed. Algorithms for other types of constraints can be developed and are not presented here for brevity. The appendix concludes with a discussion on the advantages and disadvantages of the optimality criterion methods, and their relationship to mathematical programming as applied to structural optimization.

B.2 FORMULATION

The optimization problem can be stated as:

Find the particular design vector A consisting of n elements that minimizes the objective function W(A) and satisfies the constraint equations

$$g_j(A) = C_j(A) - \overline{C}_j \leq 0 \qquad j=1,\ldots,m \tag{1}$$

where $C_j(A)$ are the functional values of the constraints, and $\overline{C}_j$ are the limiting or desired values of the constraints. When the weight of the structure is the objective function, W(A) can be written as

$$W(A) = \sum_{i=1}^{n} \rho_i A_i \ell_i \tag{2}$$

where ρ_i is the mass density, $A_i \ell_i$ is the volume of the element, A_i is the design variable and ℓ_i is a quantity associated with the geometry of the element. Equation (2) assumes that the objective function is linear in the design variables, however, this is not a requirement in developing the method. Using Equations (1) and (2), the Lagrangian can be written as

$$L(A_i, \lambda_j) = \sum_{i=1}^{n} \rho_i A_i \ell_i + \sum_{j=1}^{m} \lambda_j g_j \qquad (3)$$

where λ_j are the Lagrange multipliers. The optimality conditions are obtained by differentiating Equation (3) with respect to the design variable A_i. This gives

$$\rho_i \ell_i + \sum_{j=1}^{m} \lambda_j \frac{\partial A_i}{\partial A_i} = 0 \qquad i=1,\ldots,n \qquad (4)$$

where

$$\lambda_j g_j = 0 \qquad (5)$$

$$\lambda_j \geq 0 \qquad (6)$$

The Lagrange multipliers λ_j must be positive or zero for inequality constraints but can be negative for equality constraints. Equations (4) through (6) along with the constraint equations (Equation 1) are the Kuhn-Tucker conditions. Depending on the nature of the constraint equations, suitable optimality criterion can be obtained by using Equation (4). The n optimality criterion equations and the m constraint equations form a set of (m+n) simultaneous equations for the solution of n unknown A_i and m unknowns λ_j. Due to the nonlinear nature of these equations, they can only be solved by an iterative algorithm.

B.2.1 Displacement and Stress Constraints

When a structure is designed to satisfy multiple stress and displacement constraints, the constraint equations can be written as

$$g_j = \sum_{i=1}^{n} \frac{E_{ij}}{A_i} - \bar{C}_j \qquad (7)$$

where E_{ij} is the flexibility coefficient and is given by

$$E_{ij} = A_i \{r\}_i^t [k]_i \{s^j\}_i \qquad (8)$$

where $\{r\}_i$ is the displacement vector and $[k]_i$ is the stiffness matrix associated with the ith element. In Equation (8), $\{s^j\}_i$ is the virtual displacement vector due to the virtual loads $\{s^j\}$ acting on the structure. The nature of the virtual load depends on the type of constraint. In the case of a bar structure Equation (8) reduces to a well known relation

$$E_{ij} = \frac{F_i U_i^j \ell_i}{E_i} \tag{9}$$

where F_i is the force in the ith element due to the applied load, U_i^j is the force in the ith element due to the virtual load vector $[S^j]$, and E_i is the elastic modulus of the ith element. Differentiating Equation (7) with respect to the design variable A_i and substituting in Equation (4), the optimality criterion for a structure subjected to stress and displacement constraints can be written as

$$1 = \sum_{j=1}^{m} \lambda_j \frac{E_{ij}}{\rho_i \ell_i A_i^2} \tag{10}$$

or

$$1 = \sum_{j=1}^{m} \lambda_j \frac{e_{ij}}{\rho_i} \tag{11}$$

where e_{ij} is the virtual strain energy density in the ith element due to the jth constraint. Equation (11) states that in the optimum structure the weighted sum of the ratio of the virtual strain energy density to the mass density is equal to unity. The weighting parameters λ_j are the Lagrange multipliers.

If the virtual load system is equal to the applied load vector, then Equation (11) reduces to the criterion that in the minimum weight structure the ratio of strain energy density to mass density is the same for all the elements. This criterion is for the generalized stiffness constraint and is valid for the stress constraint problem when the maximum allowable stress is the same in all the elements.

The optimality criterion derived in this section is used to design minimum weight structures in References 3-23.

B.2.2 System Stability Constraint

The constraint equation for static buckling can be written as

$$f_j = \mu_j - \bar{\mu} \tag{12}$$

where $\bar{\mu}$ is the lowest critical buckling load factor and μ_j is the critical buckling load factor given by

$$\mu_j = - \frac{\{\phi\}^t [K] \{\phi\}}{\{\phi\} [K_G] \{\phi\}^t} \tag{13}$$

where $[\phi]$ is the buckling load, $[K]$ is the total stiffness matrix of the structure and $[K_G]$ is the geometric stiffness matrix of the structure. The buckling load of the structure is obtained by multiplying the applied load vector by μ_j. Differentiating Equation

(12) with respect to the design variable and substituting in Equation (4), the optimality criterion for a structure with stability constraints can be written as

$$1 = \sum_{j=1}^{m} \bar{\lambda}_j \frac{B_{ij}}{\rho_i A_i \ell_i} \quad (14)$$

where

$$B_{ij} = \{\phi\}_j^t [k]_i \{\phi\}_j \quad (15)$$

and

$$\bar{\lambda}_j = \frac{\lambda_j}{\{\phi\}_j^t [K_G] \{\phi\}_j} \quad (16)$$

Equation (14) can also be written as

$$1 = \sum_{j=1}^{m} \bar{\lambda}_j \frac{b_{ij}}{\rho_i} \quad (17)$$

where $b_{ij} = B_{ij}/\ell_i A_i$. This equation can be interpreted as: in the optimum structure the weighted sum of the ratio of the energy density in the buckled load to the mass density for all the elements is equal to unity. Applications of the optimization algorithm based on this criterion can be found in References 24, 25.

B.2.3 Dynamic Stiffness Constraint

The dynamic stiffness can be defined by the Rayleigh quotient and is given by

$$\omega_j^2 = \frac{\{\psi\}^t [K] \{\omega\}}{\{\omega\}^t [M] \{\omega\}} \quad (18)$$

where ω_j represents the jth natural frequency, $[\psi]$ is the natural mode of the structure and $[M]$ is the mass matrix. The constraint equation for a single frequency constraint can be written as

$$f_1 = \omega_j^2 - \bar{\omega}^2 \quad (19)$$

where $\bar{\omega}$ is the lowest required dynamic stiffness of the structure. Differentiating Equation (18) with respect to the design variable and substituting in Equation (4), the optimality criterion for dynamic stiffness can be written as

$$1 = \bar{\lambda} \frac{D_i}{\rho_i \ell_i A_i} \quad (20)$$

where

$$D_i = \{\omega\}_j^t [k]_i \{\omega\}_j - \omega_j^2 \{\omega\}_j^t [m]_i \{\omega\}_j \qquad (21)$$

and

$$\bar{\lambda} = \frac{\lambda}{\{\psi\}_j^t [M] \{\psi\}_j} \qquad (22)$$

The optimality criterion for the dynamic stiffness can also be written as

$$1 = \bar{\lambda} \frac{d_i}{\rho_i} \qquad (23)$$

where $d_i = D_i / \ell_i A_i$. This criterion states that a structure with a given frequency will be minimum weight when the ratio of the difference between the strain energy density and kinetic energy density to the mass density is the same for all the elements. The criterion is applied when designing structures with dynamic loads in References 26-27.

B.3 RECURRENCE RELATIONS

In the optimality criterion method the design variables are modified by using recurrence relations obtained from the optimality criterion. The recurrence relations also contain the Lagrange multipliers as unknowns. Therefore it is necessary to determine them before using the recurrence relations. In this section the discussion is concentrated on the optimality criterion for stress and displacement constraints as given in Equation (10). A detailed derivation of the algorithms can be found in Reference 23.

An exponential form of the recurrence relation can be written by multiplying both sides of Equation (10) by A_i^n and taking the nth root. This gives

$$A_i^{\nu+1} = A_i^{\nu} \left(\sum_{j=1}^{m} \lambda_j \frac{E_{ij}}{\rho_i \ell_i A_i^2} \right)_{\nu}^{1/n} \qquad (24)$$

where $\nu+1$ and ν are introduced to indicate the iteration numbers. In Equation (24) the parameter n controls the step size. For most structures $n=2$ gives good convergence.

A linear form of the recurrence relation can be written by expanding Equation (24) using the binomial theorem and retaining only the linear terms. This gives

$$A_i^{\nu+1} = A_i^\nu + \frac{A_i^\nu}{n}\left(\sum_{j=1}^m \lambda_j \frac{E_{ij}}{\rho_i \ell_i A_i^2} - 1\right)_\nu \tag{25}$$

In this relation the expression in parentheses is the residue in satisfying the optimality criterion. Another form of the recurrence relation can be written by transferring the expression multiplying A_i^ν in Equation (24) to the denominator and then linearizing the denominator. This gives

$$A_i^{\nu+1} = \frac{A_i^\nu}{1 - \frac{1}{n}\left(\sum_{j=1}^m \lambda_j \frac{E_{ij}}{\rho_i \ell_i A_i^2} - 1\right)} \tag{26}$$

This relation is equivalent to the linear recurrence relation obtained for the reciprocal design variable Z_i (see Reference 37). The reciprocal design variable Z_i is equal to $1/A_i$. When the optimality criterion (Equation 10) is satisfied, Equations (24) through (26) reduce to

$$A_i^{\nu+1} = A_i^\nu \tag{27}$$

indicating that it is no longer required to modify the design variables in order to improve the weight of the structure. The aim of using the recurrence relation is to modify the design variables, so that they satisfy the optimality criterion.

The recurrence relations contain the unknown Lagrange multipliers λ_j, and it is necessary to determine their magnitudes before using the relations. The Lagrange multipliers are estimated by using the condition that after the design variables are modified the active constraints are satisfied as equality constraints. If the structure is designed on the basis of a single most active constraint, an explicit expression can be derived for the Lagrange multiplier. However, in the case of a single constraint problem a precise value is not needed if the structure can be scaled to satisfy the constraints. Several methods to determine the Lagrange multipliers, with varying degrees of approximations have been proposed by different investigators. The three methods which are most suitable for the multiple constraint problem, are presented here.

A Newton-Raphson iterative method can be derived by considering a change in the constraint due to a change in the Lagrange multipliers (see Reference 10). The optimality criterion can be written as

$$A_i^2 = \sum_{j=1}^m \lambda_j \frac{E_{ij}}{\rho_i \ell_i} \tag{28}$$

The increment of the constraint Δf_j can be written as

$$\Delta f_j = \sum_{k=1}^{m} \frac{\partial f_j}{\partial \lambda_k} \Delta k \qquad (29)$$

Using the optimality criterion (Equation 28), the constraint relations (Equation 7) and Equation (29), a recurrence to determine the Lagrange multipliers can be written as

$$\{\lambda\}^{\nu+1} = \{\lambda\} - \beta [H]_\nu^{-1} \{f\}_\nu \qquad (30)$$

where $(\nu+1)$ and ν are the iteration numbers, and $[H]^{-1}$ is the inverse of the Hessian matrix where elements are given by

$$H_{jk} = -\frac{1}{2} \sum_{i=1}^{n_1} \frac{E_{ij} E_{ik}}{\rho_i \ell_i A_i^3} \qquad (31)$$

where n_1 is number of active elements. In Equation (30) the parameter is introduced to control the step size. In the Newton-Raphson method Equations (28) and (30) are used alternately to modify the design variables and estimate the Lagrange multipliers. The only disadvantage of this approach is that it is necessary to estimate the initial values of the Lagrange multipliers, and this is generally a difficult task when the number of active constraints changes from one iteration to the next.

A linear set of simultaneous equations can be derived to estimate the Lagrange multipliers by considering a change in the constraints due to a change in the design variables. The change Δf_j in the jth constraint can be written as

$$\Delta f_j = \sum_{i=1}^{n} \frac{\partial f_j}{\partial A_i} \Delta A_i \qquad (32)$$

Using the linear recurrence relation (Equation 25), and the constraint relations (Equation 7), Equation (32) becomes

$$[H]\{\lambda\}^{\nu+1} = \eta\{f\}_\nu + \{f^*\} \qquad (33)$$

where the elements of the Hessian matrix H are the same as given in Equation (31) except for the multiplier $(-1/2)$, and the elements of $\{f^*\}$ are given by

$$f_j^* = \sum_{i=1}^{n_1} \frac{E_{ij}}{A_i} \qquad (34)$$

The advantage of using Equation (33) to determine the Lagrange multipliers is that one can solve these equations without assuming any initial values. Even though Equations (30) and (33) look different, they are related to each other (see Reference 23). An equation similar to Equation (33) was proposed in Reference 6 and has been used in References 5, 17, 19, and 23.

A simple recurrent relation to determine the Lagrange multipliers can be written by assuming that all the constraints in Equation (7) are equality constraints. This gives

$$c_j = \overline{c}_j \qquad (35)$$

Multiplying both sides of this relation by λ_j^p and taking the pth root gives

$$\lambda_j^{\nu+1} = \left(\frac{c_j}{\overline{c}_j}\right)^{1/p} \lambda_j^\nu \qquad (36)$$

where (ν+1) and ν are the iteration numbers. The parameter p in Equation (36) controls the step size. Use of Equation (36) automatically eliminates passive constraints by reducing the magnitude of the corresponding Lagrange multipliers. The algorithm based on using this equation is reliable in leading to the minimum weight design. However the number of iterations required is very large, and it is necessary to assume initial values of the Lagrange multipliers. This approach is used in References 4, 5, 12, 14.

In the constraint relations given by Equation (7) we have not stipulated the gauge constraints on the sizes of the elements. Theoretically it may be necessary to include these constraints in the constraint equations and optimality criterion and determine the corresponding Lagrange multipliers. However, experience has shown that in structural optimization algorithms these constraints can be handled more efficiently by treating them as passive constraints. This eliminates the computer time required to determine the Lagrange multipliers corresponding to the gauge constraints. If any element modified by using the recurrence relation does not satisfy the gauge constraint, then the element size is changed to satisfy the constraints. Thus the elements of the structural fall into two sets. In one set the sizes are modified by the recurrence relation, and these elements must satisfy the optimality criterion at the optimum. In the other set the sizes are equal to the minimum or maximum allowable size. In the iterative algorithm the elements switch from one set to another as the design process continues. It is not necessary to know at the beginning of the design process or at any intermediate iteration exactly which elements will be active or passive at the optimum design. During each iteration one can designate a member to be active or passive depending on its status. Elements which once become passive generally remain passive during subsequent iterations. Since the design process is dynamic, even if there is an error in the initial assignment, the elements stabilize into one or the other category after a few iterations.

B.4 DISCUSSION OF DESIGN PROCEDURE

The iterative algorithm in the optimality criterion method consists of the following steps:

1) Assume initial sizes for all the elements of the structure. Normally all the elements can have the same size.

2) Analyze the structure and evaluate the stresses and displacements.

3) Scale the design to satisfy all the constraints.

4) Determine potentially active constraints.

5) Evaluate the flexibility coefficients corresponding to the potentially active constraints.

6) Determine the Lagrange multipliers corresponding to the active constraints.

7) Modify the design variables using the recurrence relation.

8) Stop the iterative procedure if the terminating criterion is satisfied or go to step 2.

In step 3, scaling the design may not be feasible for all structures. However, for those structures where it can be done, it has the advantage of obtaining a feasible design after each iteration. This also helps keep track of the improvement in the weight of the structure after each iteration.

Since the constraint which is closest to its limit is the most active, an easy method to identify active constraints is to specify a constraint range and pick up all those within the range as active constraints. The constraint range is the distance between the constraint and the constraint surface, i.e., $(C_j - \bar{C}_j)$. The flexibility coefficients E_{ij} in step 5 for the potentially active constraints can be determined by using a virtual load system. In the case of inequality constraints, the active constraint can be defined as the one which is associated with the positive Lagrange multiplier. This criterion is used in step 6 to determine the Lagrange multipliers corresponding to the active constraints and to eliminate the passive constraints. The flexibility coefficients and the Lagrange multipliers corresponding to the active constraints are used in the recurrence relation to modify the design variables.

In the iterative algorithm discussed here the number of active constraints changes from one iteration to the next. The active and passive constraints can also switch from one iteration to the next. The number of active constraints is small for the initial iterations increases with additional iterations and then stabilizes and remains the same until the optimum design is obtained.

Prior knowledge of the active constraints at the optimum is not necessary in the procedure outlined for optimality criteria method. At each iteration the set of active constraints is determined, and the same set may not remain active after modifying the design variables. The problem of determining the active constraints during the iteration is associated with all optimization methods. The minimum weight structure satisfies the optimality criterion for all the constraints, since for the active constraints the Lagrange multipliers are positive and for the passive constraints they are zero.

B.5 CONCLUSIONS AND RELATIONS TO OTHER METHODS

The algorithm, based on the optimality criterion approach discussed in the last section for the multiple constraint problem, gives a minimum weight design satisfying the theoretical optimality conditions. Obtaining a design satisfying the optimality conditions is desirable in order to check out the efficiency and the performance of the algorithm. However, in a practical design problem the use of such a design may not be feasible due to manufacturing considerations. In many practical cases it is more useful to obtain a near minimum weight design with a good distribution of material than a design satisfying the theoretical optimality conditions. A near optimum design for certain types of problems can be obtained by designing the structure only on the basis of a single or a few dominant active constraints and treating the other constraints as passive. A good example of this is the design of a structure with displacement and stress constraints. In this case if the displacement constraints are dominant, then one can obtain a nearly minimum weight design by considering only the displacements in the constraint equations and treating the stress constraints as passive constraints. The elements violating the stress constraints can be scaled to satisfy the maximum allowable stress in that element. The stress constraints are then not explicitly included in Equation (7). This simplification saves substantial computer time in determining the flexibility coefficients and the Lagrange multipliers. This and similar approximations, though they have been used in the development of the optimality criterion based algorithms, should not be construed as essential in the development of these methods. These approximations simplify the algorithms and make them more efficient with respect to the use of computer time. The algorithms based on these approximations may not give an optimum design but give a near optimum design with less computational effort. This approach may be essential and desirable when the structure is idealized by thousands of elements.

Recently some investigators have preferred to use the repiprocal design variable $Z_i(=1/A_i)$ instead of the direct design variable A_i. The effect of this change in the definition of the problem on the algorithm based on the optimality criterion is discussed in Reference 37. It is shown that the optimality criterion (Equations 10, 28), the exponential recurrence relations (Equation 24), and the equations used to determine the Lagrange multipliers (Equations 30, 33, 36), are not affected by changing the design variables. These relations and equations expressed in terms of A_i or Z_i can be obtained from one another by using the relationship between A_i and Z_i. Since the analysis and the evaluation of flexibility coefficients are not affected by the definition of the problem in terms of A_i or Z_i, there is no specific advantage in defining the problem in terms of Z_i. The linear recurrence relation in terms of Z_i however is not equivalent to Equation (25), but it is equivalent to Equation (26). Even with this difference Equation (33) used to evaluate the Lagrange multipliers remains unchanged (see Reference 37).

The behavior of the algorithm depends primarily on 1) the recurrent relation or search formula used to modify the design variables during the iterations and 2) the equations used to evaluate the Lagrange multipliers. The generality or rigor used in the definition of the problem, the scheme selected to determine the active and passive constraints, or the method used to find the step size have generally a secondary effect on the efficiency of the algorithm. If an identical recurrence relation to modify the design variables and equations to evaluate the Lagrange multipliers can be derived by a method other than the optimality criterion method, the behavior of the algorithm based on this method will be the same. In Reference 37 it is shown that the linear recurrence relations for the design variables A_i and the reciprocal design variables Z_i, and the corresponding equations to determine the Lagrange multipliers obtained by the optimality criterion approach can be derived by the projection method of nonlinear programming. An algorithm based on the dual method of mathematical programming is developed in Reference 38. In this reference the problem is defined in terms of the reciprocal design variable. The search formula obtained in this method is equivalent to Equation 28, and the Lagrange multipliers are calculated by using Equation 30, where the step size parameter β is determined so as to maximize the Lagrangian as a function of the Lagrange multipliers.

REFERENCES

1. Prager, W. and Taylor, J. E., "Problems of Optimal Structural Design," J. Appl. Mech. Trans. ASME, 35, pp. 102-106, 1968.

2. Sheu, C. Y. and Prager, W., "Recent Developments in Optimal Structural Design," Appl. Mech. Rev. 21, pp. 955-992, 1968.

3. Venkayya, V. B., Khot, N. S., and Berke, L., "Application of Optimality Criteria Approaches to Automated Design of Large Practical Structures," Second Symp. Struct. Opt., AGARD-CP-123, Milan, Italy, 1973.

4. Berke, L. and Khot, N. S., "Use of Optimality Criteria Methods for Large Scale Systems," AGARD Lecture Series No. 70, Struct. Opt., 1974.

5. Khot, N. S., Venkayya, V. B., and Berke, L., "Experience with Minimum Weight Design of Structures Using Optimality Criteria Methods," Second Int. Conf. Vehicle Struct. Mech., Southfield, Michigan, 1977.

6. Kiusalaas, J., "Minimum Weight Design of Structures via Optimality Criteria," NASA TN D-7115, 1972.

7. Venkayya, V. B., Khot, N. S., and Reddy, V. S., "Energy Distribution in an Optimum Structural Design," AFFDL-TR-68-156, 1969.

8. Berke, L., "An Efficient Approach to the Minimum Weight Design of Deflection Limited Structures," AFFDL-TM-70-4-FDTR, 1970.

9. Venkayya, V. B., "Design of Optimum Structures," J. Computers and Struct. 1, pp. 265-309, 1971.

10. Taig, I. C. and Kerr, R. I., "Optimization of Aircraft Structures with Multiple Stiffness Requirements," Second Symp. Struct. Opt., AGARD-CP-123, Milan, Italy, 1973.

11. Gellatly, R. A. and Berke, L., "Optimality Criterion Algorithms," Optimum Structural Design, edited by R. M. Gallagher and O. C. Zienkiewicz, John Wiley and Sons, London, 1973.

12. Nagteggal, J. C., "A New Approach to Optimal Design of Elastic Structures," Computational Methods in Applied Mechanics and Engineering, Vol. 2, pp. 255-264, February 1973.

13. Khot, N. S., Venkayya, V. B., Johnson, C. D., and Tischler, V. A., "Optimization of Fiber Reinforced Composite Structures," Int. J. Solids Structures, Vol. 9, pp. 1225-1236, Pergamon Press, September 1973.

14. Khot, N. S., Venkayya, V. B., and Berke, L., "Optimum Design of Composite Structures with Stress and Deflection Constraints," AIAA Paper No. 75-141, presented at AIAA 13th Aerospace Sciences Meeting, Pasadena, California, 1975.

15. Gellatly, R. A. and Berke, L., "Optimal Structural Design," AFFDL-TR-70-165, Air Force Flight Dynamics Laboratory, Wright-Patterson AFB, Ohio, 1975.

16. Gorzynski, J. W. and Thornton, W. A., "Variable Energy Ratio Method for Structural Design," J. Struct. Div. ASCE, 101, No. ST4, pp. 975-990, 1975.

17. Rizzi, D., "Optimization of Multi-Constrained Structures Based on Optimality Criteria," Proc. AIAA/ASME/SAE 17th Struct., Structual Dynamics and Materials Conf., King of Prussia, Pennsylvania, pp. 448-462, 1976.

18. Dobbs, M. W. and Nelson, R. B., "Application of Optimality Criteria to Automated Structural Design," AIAA J. 14, pp. 1436-1443, 1976.

19. Segenreich, S. A., Zouain, N. A., and Herskovits, J., "An Optimality Criteria Method Based on Slack Variables Concept for Large Scale Structural Optimization," Proc. Symp. Applications of Computer Methods in Engineering, (Ed. C. Wellford, Jr.), University of Southern California, pp. 563-575, 1977.

20. Austin, F., "A Rapid Optimization Procedure for Structures Subjected to Multiple Constraints," Paper No. 77-374, AIAA/ASME/SAE 18th Struct., Structural Dynamics and Materials Conf., San Diego, California, 1977.

21. Rao, G. V., Shore, C. P., and Narayanaswami, R., "An Optimality Criterion for Resizing Heated Structures with Temperature Constraints," NASA TN D-8525, 1977.

22. Khah, M. R., Willmert, K. D., and Thornton, W. A., "A New Optimality Criterion Method for Large Scale Structures," AIAA/ASME 19th SDM Conf., Bethesda, Maryland, pp. 47-58, 1978.

23. Khot, N. S., Berke, L., and Venkayya, V. B., "Comparisons of Optimality Criteria Algorithms for Minimum Weight Design of Structures," AIAA/ASME 19th SDM Conf., Bethesda, Maryland, pp. 37-46, 1978.

24. Kiusalaas, J., "Optimum Design of Structures with Buckling Constraints," Int. J. Solids and Struct. 9, pp. 863-878, 1973.

25. Khot, N. S., Venkayya, V. B., and Berke, L., "Optimum Structural Design with Stability Constraints," Int. J. Numerical Methods Engineering 10, pp. 1097-1114, 1976.

26. Venkayya, V. B., Khot, N. S., Tischler, V. A., and Taylor, R. F., "Design of Optimum Structures for Dynamic Loads," Third Conf.

Matrix Meth. Struct. Mech., Wright-Patterson AFB, Ohio, pp. 619-658, 1971.

27. Venkayya, V. B. and Khot, N. S., "Design of Optimum Structures to Impulse Type Loading," AIAA J. 13, pp. 989-994, 1975.

28. Segenreich, S. A. and McIntosh, S. C., "Weight Minimization of Structures for Fixed Flutter Speed via an Optimality Criterion," Proc. AIAA/ASME/SAE 16th Struct., Structural Dynamics and Materials Conference, Denver, Colorado, 1975.

29. Austin, F., et al., "Aeroelastic Tailoring of Advanced Composite Lifting Surfaces in Preliminary Design," Proc. AIAA/ASME/SAE 17th Struct., Structural Dynamics and Materials Conf., Valley Forge, Pennsylvania, 1976.

30. Wilkinson, K., Lerner, E., and Taylor, R. F., "Practical Design of Minimum-Weight Aircraft Structures for Strength and Flutter Requirements," J. Aircraft, 13, pp. 614-624, 1976.

31. Kiusalaas, J. and Shaw, R., "An Algorithm for Optimal Design with Frequency Constraints," Int. J. for Numerical Methods in Engineering, Vol. 13, No. 2, December, 1978.

32. Gellatly, R. A., Dupree, D. M., and Berke, L., "OPTIM II: A Magic Compatible Large Scale Automated Minimum Weight Design Program," AFFDL-TR-75-97, I and II, 1974.

33. Khot, N. S., "Computer Program (OPTCOMP) for Optimization of Composite Structures for Minimum Weight Design," AFFDL-TR-76-149, 1977.

34. Venkayya, V. B. and Tischler, V. A., "OPTSTAT - A Computer Program for Optimal Design of Structures Subjected to Static Loads," AFFDL-TR-79-(in preparation).

35. Kiusalaas, J. and Reddy, G. B., "DESAP 2 - A Structural Design Program with Stress and Buckling Constraints," NASA CR-2797 to 2799 (3 volumes), National Aeronautics and Space Administration, Washington, D.C., 1977.

36. Venkayya, V. B., "Structural Optimization: A Review and Some Recommendations," Int. J. for Numerical Methods in Engineering, Vol. 13, No. 2, December 1978.

37. Khot, N. S., Berke, L., and Venkayya, V. B., "Minimum Weight Design of Structures by the Optimality Criterion and Projection Method." AIAA 79-0720, paper presented at AIAA/ASME/ASCE/AMS 20th Structures, Structural Dynamics and Material Conference, St. Louis, Missouri, April 1979.

38. Schmit, L. A., Jr. and Fleury, C., "An Improved Analysis/Synthesis Capability Based on Dual Methods - ACCESS 3," AIAA Paper 79-0721, presented at 20th Structures, Structural Dynamics and Materials Conference, St. Louis, Missouri, April 1979.

RAYMOND H. FOGLER LIBRARY
DATE DUE

BOOKS ARE SUBJECT TO
RECALL AFTER TWO WEEKS